建筑立场系列丛书 No.44
节能与可持续性
Energy efficient, Sustainable
中文版
（韩语版第360期）
韩国C3出版公社 编
张琳娜 周一 译
大连理工大学出版社
U0193747

C3 建筑立场系列丛书 No. 44

12

节能与可持续性
——迈向绿色未来的关键一步

16

能源构建未来

92

七抹绿荫

C3 No. 44 Energy Efficient and Sustainable

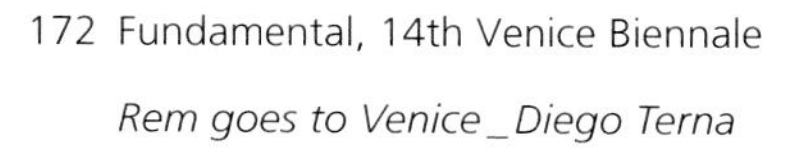

Critical Step towards Green Future

Energy efficient and Sustainable

Energy builds the future

Seven Shades of Green

金门客轮码头竞赛作品_Junya Ishigami+Associates

金门县客轮码头的竞赛作品评选产生了五个获奖团队。其中，石上纯也当选为冠军。

获奖项目是为金门岛上的客轮码头设计的方案，金门岛是台湾管辖的一个小岛，距离台湾仅一小时的飞行航程。

许多具有地方风格的村庄是当地宝贵的历史文化遗产。这些村庄具有特色的屋顶成行排列，形成了令人难忘的景观。

这个小岛曾在历史上的不同时期成为战场，许多地方仍然残留着战争的遗迹。

此外，岛上的自然环境不适宜植被生长，而且，自17世纪以来森林的过度采伐导致岛上多年不见绿色植物。然而，自1950年左右开始，由于战争的爆发，为了隐藏士兵和军事设施，岛上实施了大规模的植树造林活动。如今，岛上已是一片碧绿，并被指定为国家森林公园。

文化、战争和自然三者间的复杂关系造就了岛屿的独特性。当今，由于岛上丰富的自然和人文资源，旅游成为岛上的主要产业。为了适应在不久的将来持续增长的游客数量的需要，一项大型客运码头的规划应运而生。该区域面积广阔，约100m宽，520m长，由金门岛海岸线上的垃圾填埋场形成。包括停车区在内，整座建筑区域占地约60 000m²。

建筑师说："在这座独特的小岛上，我们想要创造一个给人以崭新和光明的形象的环境，作为金门岛价值的延伸。我们想要打造一个既本土化，同时又具有普遍性，且人人可以共享的项目。

我认为屋顶的外形非常接近于山峰。

屋顶成行排列形成的风景恰似群山之地。建筑就是一座小山峰。看着岛上这些独特的、极具地方特色的屋顶，我在想是否可以将它们看做是成就了风景的山峰，是否可能将一座大型建筑结构设计成为一处新的自然环境。

这处人工景观与岛上的自然遗产之间存在着一定的联系。"

Port of Kinmen Ferry Terminal Competition

The international competition for Kinmen County has resulted in five winning teams. Among them, Junya Ishigami has been selected as the first prize winner.

The winning project is a plan for a passenger ship terminal on a small island called Kinmen, Taiwanese territory located at a distance of about one hour flight from Taiwan, China.

Historic cultural heritage of the numerous villages built in is of vernacular style. These villages are marked with characteristic roofs, the repetition of which creates a memorable landscape.

The land was a battlefield in various stages of history, where many remnants of the war still remain.

项目名称：Port of Kinmen Passenger Service Center
地点：Kinmen County, Taiwan
建筑师：Junya Ishigami+Associates
当地建筑师：Bio Architecture Formosana
顾问：Ove Arup&Partners
甲方：Kinmen County Government
用途：Passenger Ship Terminal
用地面积：52,000m²
总建筑面积：26,000m²
有效楼层面积：approx. 65,000m²
设计时间：2014—2015
施工时间：2015—2017

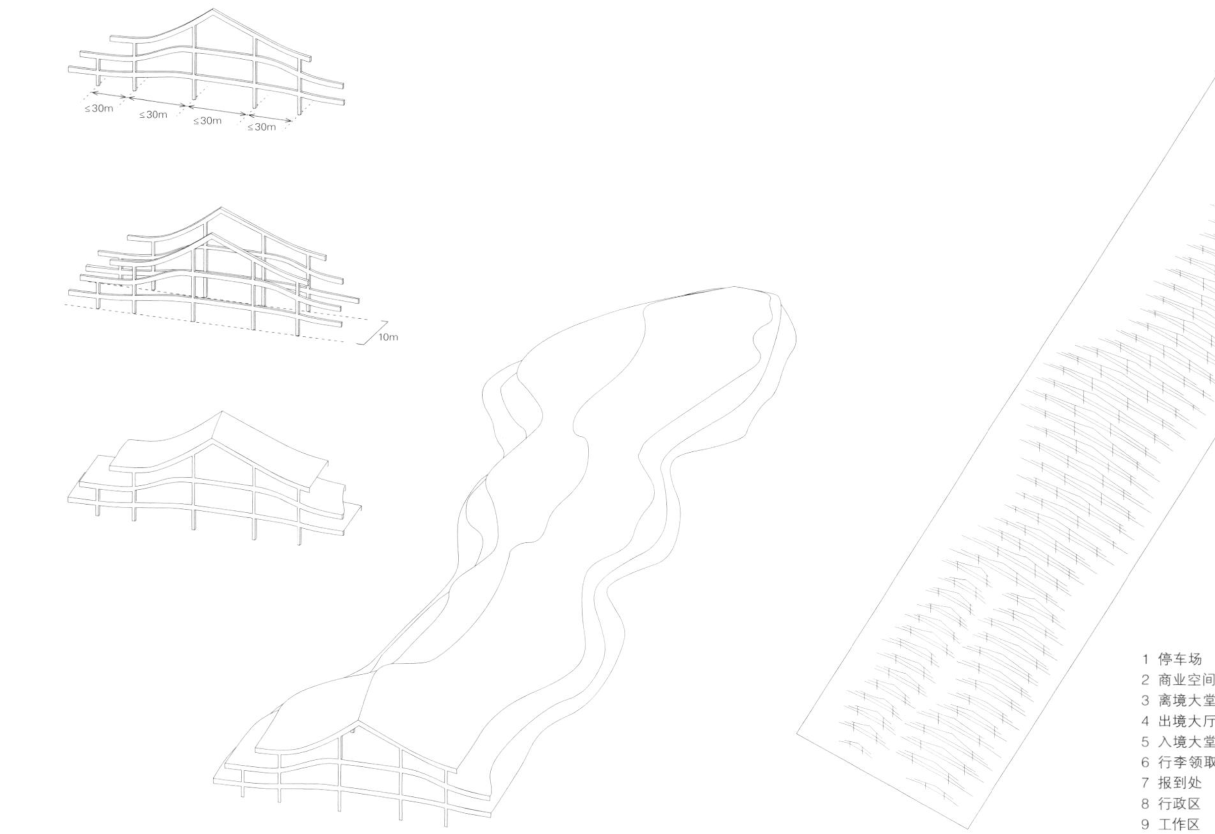

A-A' 剖面图 section A-A'

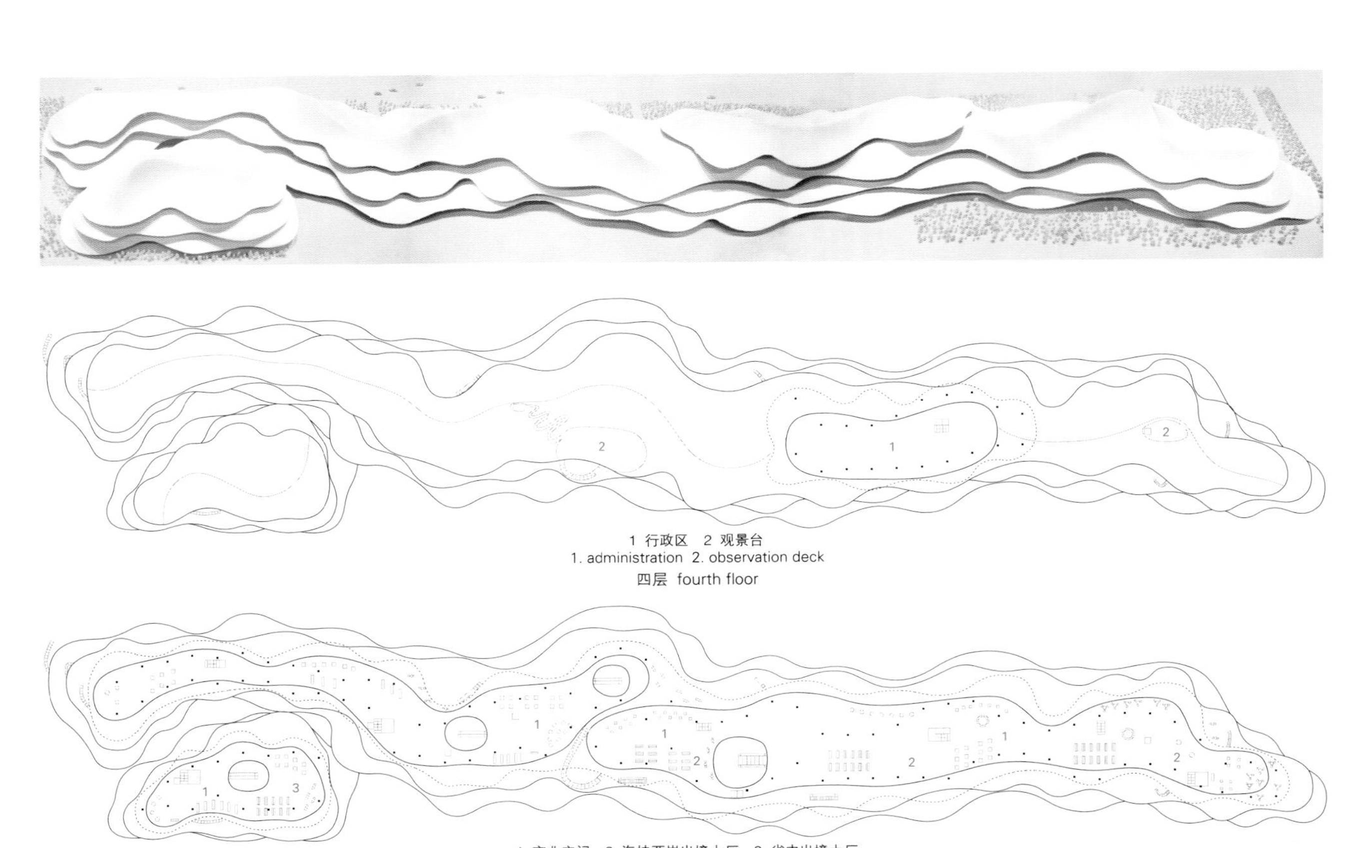

1 行政区 2 观景台
1. administration 2. observation deck
四层 fourth floor

1 商业空间 2 海峡两岸出境大厅 3 省内出境大厅
1. commercial space 2. cross-strait departure lobby 3. domestic departure lobby
三层 third floor

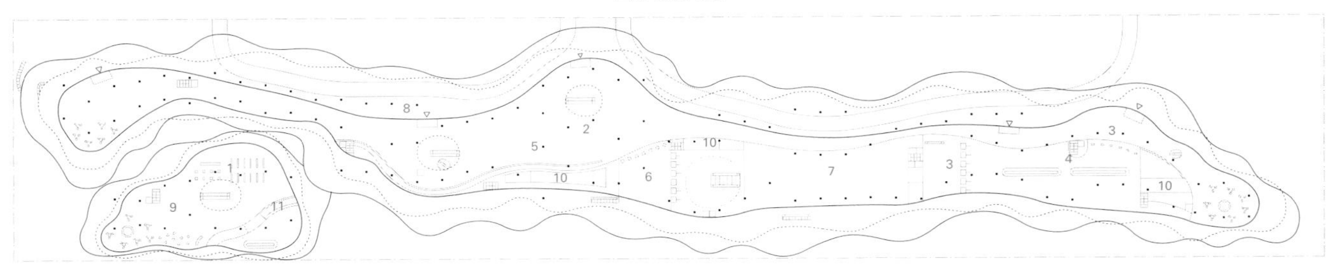

1 商业空间 2 海峡两岸离境大堂 3 海峡两岸入境大堂 4 海峡两岸行李领取处 5 海峡两岸报到处
6 护照检查处 7 行政区 8 下车区 9 省内离境大堂 10 旅客服务空间 11 省内出境大厅
1. commercial space 2. cross-strait departure hall 3. cross-strait arrival hall 4. cross-strait baggage claim 5. cross-strait check-in counter
6. passport control 7. administration 8. drop-off area 9. domestic departure hall 10. traveler service space 11. domestic departure lobby
二层 second floor

1 搭车区 2 下车区 3 工作区
1. pick-up area 2. drop-off area 3. work space
一层 first floor

In addition to the island's naturally adverse environment for vegetational growth, excessive deforestation beginning in the 17th century resulted in years of almost no greenery. However, triggered by the war beginning around 1950, large-scale afforestation was carried out to conceal the soldiers and military installations. Now, the land is rich in verdure and is designated as a national park.
Culture, war, and nature. The complex relationships among these elements are what make this island unique. Currently, based on its numerous assets, the island's main industry is tourism. To accommodate the increasing number of tourists in the near future, a plan for a large passenger ship terminal has been initiated. The site is extensive, measuring 100m on the short side and 520m on the long side, and is formed by a landfill on the Coast of Kinmen Island. The total area of the building is approximately 60,000m² including the parking space.
The architect said *"On this small and unique island, we thought to create an environment with a new and bright impression as an extension of the values of this place. We want to create something native, yet at the same time universal, something that can be shared by everyone.*
I think roofs are very similar to mountains. The scenery in which roofs arrange themselves resembles mountainous regions. Architecture is a small mountain.
By looking at the island's unique and indigenous roofs, I wonder if it is possible to see them as something akin to mountains, creating a scenery, if it is possible to plan a massive structure as a new natural environment. This artificial natural environment bears a certain link to the island's natural legacy."

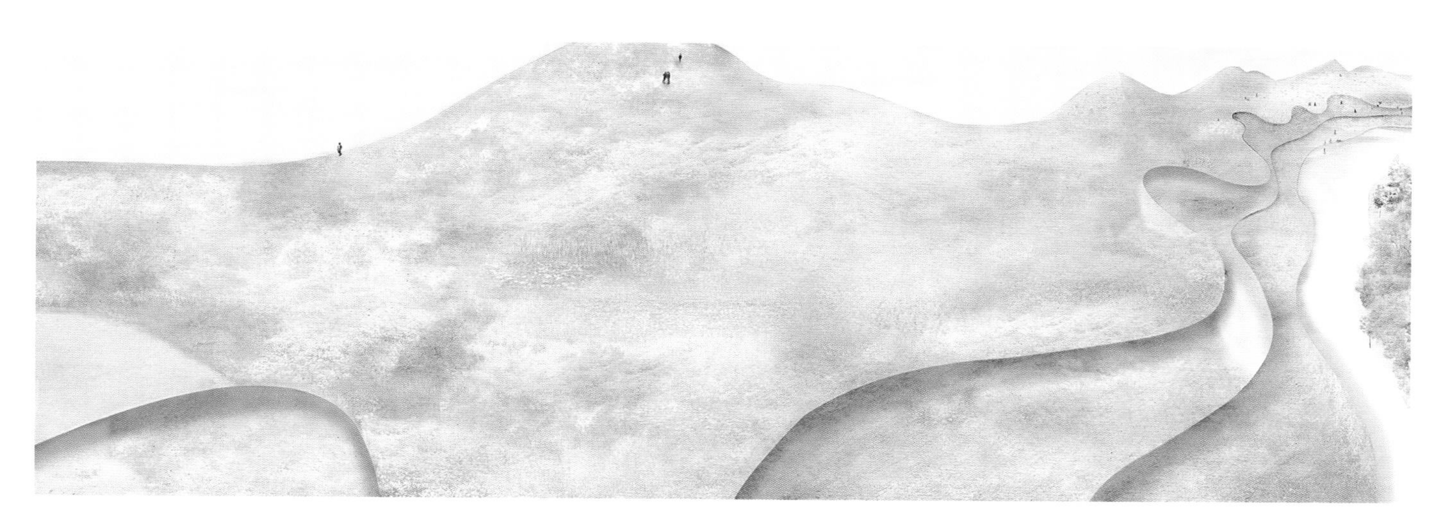

鲁汶天主教大学新建筑学院设计比赛_Aires Mateus

艾瑞斯·马特乌斯赢得了鲁汶天主教大学新建筑学院的设计比赛。

图尔奈市（比利时的西南部城市）由于过去多次被用作建筑案例，因此得到了人们的认可和重视。这座城市以及它的都市风貌都受到了各种社会动荡因素的影响，但它今天的繁荣和经历过的动荡却有着密不可分的关系。在整个大扩张和大沉睡时代，这座城市吸收并保留了不同时代的历史痕迹。毋容置疑，这座城市的纪念性在象征性建筑中得以清晰的表达。

新建筑从限制中产生，从创造一种全新的建筑价值的决心中解脱。"公共空间"在设计图纸中是考虑的首要因素，其体量是相邻建筑物直接产生的结果。公共空间会很自然地"流"入新建筑内。而新结构不会对原有建筑物产生任何接触，除了通道。不过仍然有一条光线缝隙增强了不同时代的建筑物之间的联系。从建成体量（由其边界来确定）来看，原始的形式有所改变。所有被移除的空间都参考了传统意义上属于我们的公共遗产的形式。这些空间不仅唤醒了赋予图尔奈特色的通道，也唤醒了建筑师建立的与庇护所或者是入口处的连接。

建筑的边界以这种方式在其内部围出一个空体量。这处大型室内空间被设计成一个大型的上空空间。正如大家所期待的，它被用于承载全体教员所发生的不可预测的生活事件。人们在LOCI街区的各个方向都能看到中殿，正如所期望的图尔奈大教堂的空旷性一样。

从立面来看，室内空间的原型被拆除，以形成模具。它们成为新建筑公共区域里不可缺少的一部分。这里是供人们在各个通道之间聚集并交流的场所。它不仅仅是所有街区的主干道，也是人们进行深入沟通以及共生合作的生活中心。这是一个可以接收城市并使其自身充满可用性的中央场地。这里可以是一个集合地点，一个社交场所，或者只是区域中心内的一条简简单单的人行道。

修建此处的意图在于致力于推动城市生活，并进行更加深入的文化交流。最大的广场内有一小片种植着当地树种的树林，它们使这里转化为理想的生活区。

图尔奈遗址里安详宁静的土地已经被唤醒，并且在之前从未存在的区域内规划了一个连接点。

Catholic University of Louvain, Architecture School Competition

Aires Mateus has won a competition to design the new architecture school of Catholic University of Louvain.

The importance of Tournai is recognized and distinguished through countless architectural examples built over time. The city, as well as its urbanity, was subjected to various oscillations of its importance proportional to its prosperity. Among the periods of greatest expansion and of greater lethargy, the city has adapted and retained traces of these different periods. There is an indubitable monumentality, expressed in clear iconographical buildings.

The new building emerges from the constraints, but especially from the ambition to establish a new architectural value. The "public space" is elected as a priority for the drawing. The volume is a direct result of the contiguous buildings. Their dimensions are poured into the new building.

项目名称：Faculté d'Architecture de Tournai
地点：Tournai, Belgium
建筑师：Aires Mateus
协调建筑师：Jorge P Silva
合作建筑师：Sara Nobre, Sofia Paradela, Inês Gulbenkian, Bernardo Sousa
施工单位：Jean Pierre Beuls, Stéphane Delpire
甲方：Université Catholique de Louvain
用地面积：5,990m²
施工面积：7,010m²
造价：EUR 4,130,000
施工时间：2014
竣工时间（预期）：2015

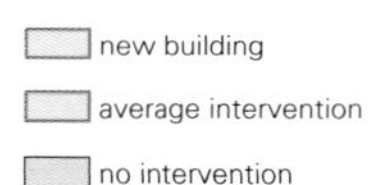

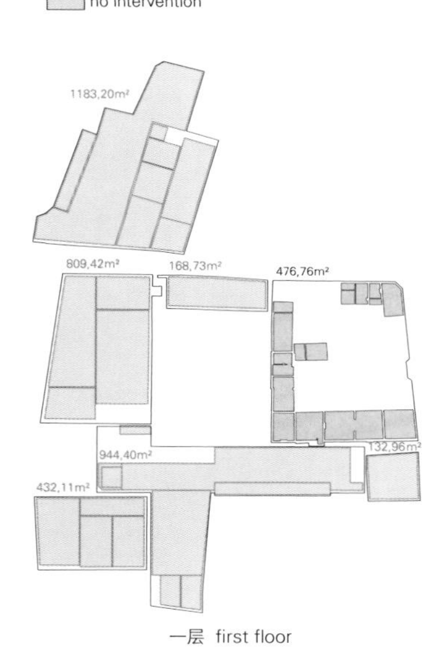

一层 first floor

二层 second floor

三层 third floor

The new construction does not touch the existing buildings except for passages. There's a gap of light that reinforces the relationship between buildings from different eras.
From the built volume that is defined by its perimeter, archetypal forms are subtracted. All this "air" that is removed refers to the forms that traditionally belong to our common heritage. These are evocations, not only of the passages that characterize Tournai, but also of the connection that the architects establish with their own idea of a shelter, or preparation for entry.
The perimeter encloses in this way an internal hollow volume. This large interior space is designed as a huge void. It is a volume of air in expectation, capable of containing life bearing the unpredictability of a Faculty. A central nave that can be foreseen from all sides of the LOCI Block, the same way the hollowness of the Tournai Cathedral is anticipated.
From the facades archetypes are subtracted, producing their counter mold in the interior space. Their presence becomes part of the common areas in the new construction. This is the space of communion and gathering between all the passage fluxes. This is not only the spine of all the blocks, but also the center of life that makes all the context communicate and live in symbiosis. This is the central space which receives the city and makes itself available for it. It can serve as a venue, a socializing area, or a simple passage in the heart of the quarter.
It is intended that the new quarter is dedicated to urban life, and more deeply to cultural exchange. On the largest square a small grove of native trees is created that can transform this space into a living area. The serene presence of Tournai's heritage is evoked, and proposes a link within the quarter which had never existed.

东南立面 south-east elevation

西南立面 south-west elevtation

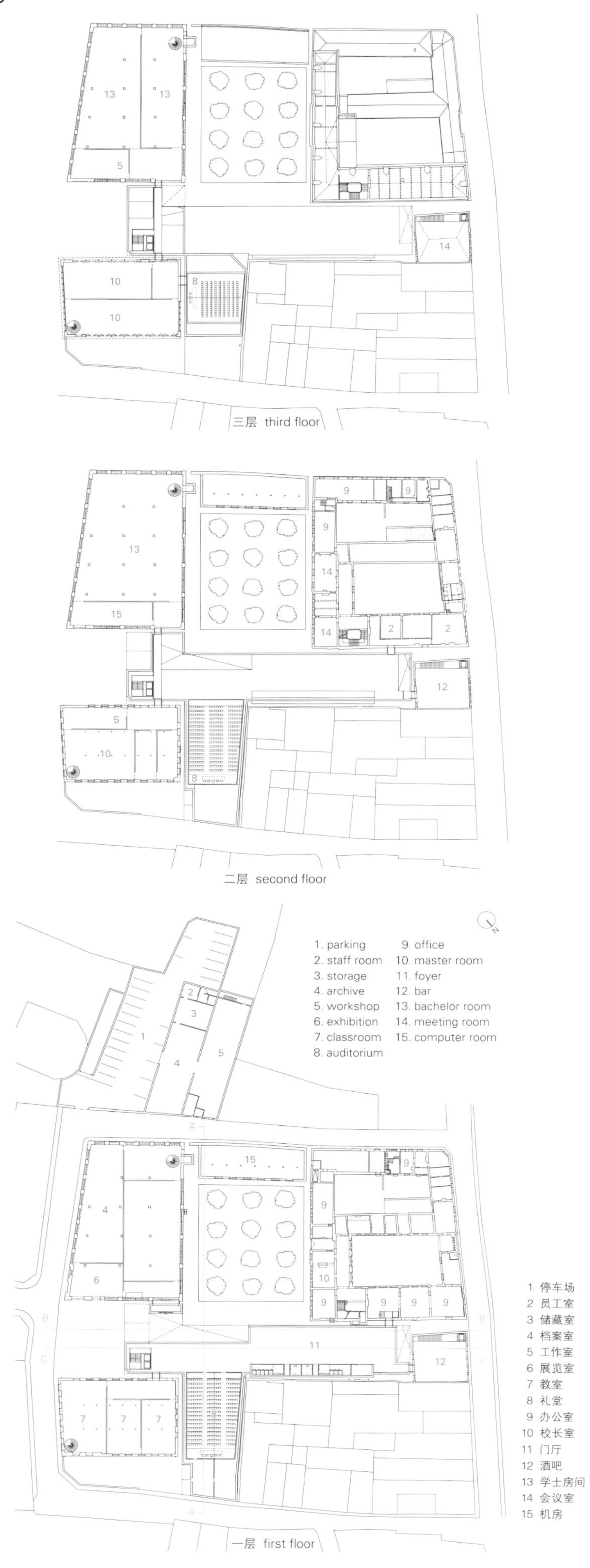

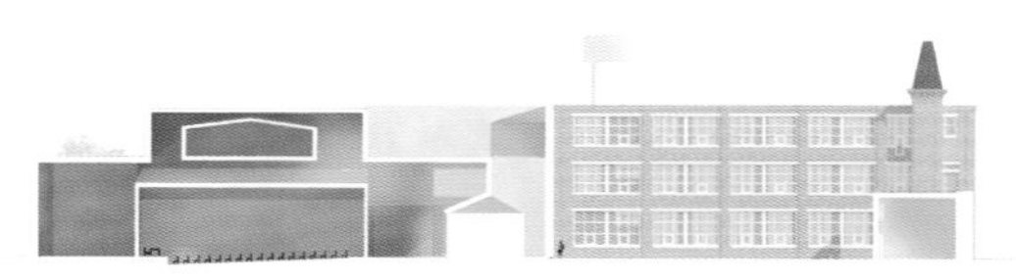

A–A' 剖面图 section A-A'

B–B' 剖面图 section B-B'

C–C' 剖面图 section C-C'

节能与可持续性

迈向绿色未来的关键一步

Energy efficient 'n Sustainable

Critical Step towards Green Future

关于气候变化的讨论主要集中在化学排放和化合物如上层大气中的二氧化碳浓度上升的危险方面。反过来，这通常又与内燃发动机和塑料制品大量消耗石油化工产品密切相关。然而这是一个过于简单的想法，因为它通过与快速而短期的产品消耗之间的关联，将问题限制在了经济市场中的运输和制造业这一部分。这一观点简单地排除了建筑对于气候变化的责任。这是一个让人不愿接受的事情，因为我们都在其中发挥作用，且要在一定程度上反思自己的行为和生活方式，建筑虽然对环境没有即时的、明显的影响，但也因其长久的使用年限而起到一定的作用。在一天或是一周这样一个很短的时期内，建筑对于环境的影响是极小的，但是在其一生、通常是几百年的时间里所造成的影响将是巨大的。如果考虑建筑的社会和文化内涵，那么规划师和建筑师的责任将更加重大。我们为了我们这一代人的需要建造楼房，但这些建筑也将被我们的孙辈使用。如果考虑长期的影响，建筑对于环境所造成的伤害将紧随汽车之后。虽然"可持续发展"这个词隐含了现代人的一个担忧，或者一时解答了Ann Thorpe[1]提出的第一个问题，但是还有另外一个问题，即"什么是我们努力想要维持下去的？"

退一步从全球性视角来看这个问题，我们不禁要问：关于建筑我们不惜耗费如此多的精力和讨论的目的是什么？所有规划的建筑（我们不得不承认，其中只有大约百分之十是由建筑师设计的[2]）都有自身固有的需求层次，无论是纯粹的基本功能性庇护所还是市政大楼。如今，建筑师需要在背景和贯穿整体的环境理解下来设定所需的参数。这些也许是物质上的影响，而此时这个供应链与建筑的实际容量、能源或水的供给和消耗同等重要。

The majority of discussion on Climate Change is focused on chemical emissions and the dangerous increase in concentration of compounds such as carbon dioxide in the upper atmosphere. This in turn is normally associated with the consumption of petro-chemicals, simply put by the internal combustion engine and plastics. However this is a simplistic approach as its ring fences the issues within the economic market's sectors of transportation and manufacturing indeed by associating with the consumption of products which are transient and short term. This viewpoint conveniently places the responsibility of climate change elsewhere from architecture. This is a distraction as we all have a role to play and a level of introspective questioning about our own practice and lifestyles is required. Buildings, although without such an immediate and explicit impact on the environment, do form part of the debate due to their longevity. Impacts over a short term, a day or week are minimal but with a lifetime often in the hundreds of years, the consequences can be great. If we include social or a cultural implications of buildings then the responsibility of planners and architects becomes even greater. We are creating buildings for our needs but which will be used by our grandchildren. If we include impacts over time then buildings elevate themselves to key villains to the environment along with cars. Although the word "sustainability" has the connotation of a modern concern or the first question raised by Ann Thorpe[1], the other is "What are we trying to sustain?"
Stepping back from the global picture we need to ask what is the purpose of the building to which we expend so much energy and discussion? All planned buildings, and here we have to acknowledge that only approximately 10% of all buildings are designed by architects[2], have their own intrinsic hierarchy of need from the purely functional basic shelters to our Municipal Edifices. Architects now need to frame these parameters of needs within a backdrop and over-arching understanding of environmental impacts. These may be material impacts and here the supply chain is equally important as the physical bulk, energy or water supply and consumption.

我们的目标是将对环境的影响降低为零。为什么？答案有两方面。首先要允许整个星球和它支持的所有生态系统（我们作为人类是否真的想要减少甚至消灭其他物种？）保持平衡，其次要允许未来的几代人能够有机会享有与我们相同的选择权。最初，我们拒绝承认气候变暖，并且对于成功的判断标准感到迷惑。部分原因是受到错误的观念所驱使，认为追究生态影响的代价将是困难和昂贵的。许多人甚至将“可持续发展”看作是一种负担或者另外一个需要克服的困难。但下面的章节中所展示的范例则证明了环境的影响可以减小。环境问题再次触动了贯穿建筑的创造性思维的活力。这不仅产生了良好的环境反应，同时还通过感知建筑、欣赏建筑以及与建筑交流的方式，来形成一种模式转移。

我们需要将我们所说的可持续发展和环境问题拆开来看。作为回应，我们通过判定方法和量化解决方案来进行估测。简单的观察与开明的回应相匹配，进而形成我们推动的目标。能源问题与消耗的体量相关。我们能减少这一问题所带来的影响吗？我们需要从无法替代的原始材料中提取能源吗？亦或者，转换能源能加重环境破坏吗？能源运输也并非是高效的，那么我们需要重新考虑我们输送的能源的规模吗？我们能否在本地生产能源？然而具有讽刺意味的是，在每一天，这个星球表面，有比我们使用的更多的能源在以日照的方式进行输送。只是现在我们还没有足够的技术来获取这类能源。我们现在仍然采用西方世界的视角，即自工业革命以来，技术的强力能够解决所有的问题。将自然看做是灵感与典范的科学家们注意到地球与自然的韵律并不具有破坏性，而是能够形成一种平衡，来获取替代能源。依赖化石能源在过去只是权宜之计，但是在未来却不是最佳的解决方案。这也因此产生了两个影响，首先是对能源来源的依赖，而这些能源终将被

Our goal is to reduce the impact on the environment to zero. Why? The answer is twofold, firstly to allow the planet and all the ecosystems it supports (do we as humans really want to reduce or even eliminate the populations of other species?) to remain in balance and secondly to allow future generations to have the same choices that we are privileged to have. Initially there has been both denial of Climate Change and confusion on the criteria by which we judge success. This has in part been driven by a perception that addressing ecological impacts will be both difficult and costly. Many even considered “sustainability” as a burden or another difficulty to overcome. But the exemplars in the following chapters demonstrate, impacts can be reduced. Environmental issues have reinvigorated creative thinking through buildings. This has created not only good environmental responses but a paradigm shift in the way we conceive, appreciate and interact with buildings.

We need to unpick what we mean by sustainability and environmental issues. In response we can set gauges by which we can both justify approaches and quantify solutions. Simple observations matched with enlightened responses achieve objectives in ways we can celebrate. Energy issues are both about volume of consumption, can we reduce the problem? And do we need to take energy from sources which cannot be replaced? Or by the conversion to energy do we increase environmental damage? Transportation of energy is also in-efficient, do we need to reconsider the scale of what we deliver? Can we generate energy locally? Ironically there is more energy than we can use delivered across the planet's surface each day in the form of sunshine. We are just not good enough yet at harvesting this energy. We are still mentally locked into the western world's viewpoint, from the Industrial Revolution, that the brute force of technology will solve all problems. Scientists who look at nature for inspiration and examples note that the rhythms of the earth and nature are not destructive but create balance, and harvest what can be replaced. Dependency on fossil fuel was expedient in the past but not a solution for the future. This has two impacts, firstly the dependence on a source of energy will eventually run out. Secondly the conversion of this stored energy to us-

消耗掉。其次是将储存能源转换为可用能源，这种转换关乎物质从一种形式转化为另外一种形式，也关乎从地球上的某一个点（经常指地下）传送到另外一个点（地上）。Karl-Henrik Robert和The Natural Step（TNS）公司便将后者视为一个问题，并且声明要将材料保存在原有场地。

创造性并非是绝对的事物，而边界也并非是清晰的。在可持续性的范围内，仍然存在着关于文化和社会建筑的问题。建筑以及建筑师所定义的室内和室外空间的设计对我们构建社会来说是十分关键的，包括微观社会（一个家庭）和宏观社会（一座小镇）。这便要求建筑师和规划师投入更多。将建筑视为一个纯粹的物理实体的观念是错误的，建筑的物理方面可能带来环境方面的影响，但是一处规划得体的空间或者建筑能够增强社会凝聚力和互动性。在这一方面，建筑成为我们最重要的产品。实际上，建筑师和规划师能够创建鼓励新型社交和商业活动的空间。那么在现在和未来，我们要怎样生活？建筑师成为这一问题的中心要素。

Alastair Fuad-luke[3]则从可持续性的整体性角度支持幸福安康是连接环境与人类、经济和未来的要素这一理念。它使建筑成为一个特权、一个富有创意的义务。接下来所介绍的案例会重点突出建筑师以模拟的方式来应对挑战，以降低影响，增强我们与建筑彼此之间的关系。

able energy converts matter from one form into another and also from one point on the planet(usually under the ground) to another (in the upper atmosphere). Karl-Henrik Robert and The Natural Step(TNS) highlight the latter as a problem with almost a manifesto for keeping materials in the original place.

Creativity does not deal in absolutes and boundaries are blurred. Within sustainability there are questions about culture and social constructs. The design of both buildings and the internal and external spaces which they define is critical to the way we organize societies. This goes from the micro(the home) to the macro(a town). This demands more rather than less input from architects and planners. To view a building as a purely physical entity is wrong. The physical aspects of a building may cause the environmental impacts but a well planned space or building enhances social cohesion and interaction. In this aspect buildings become our most important products. In deed architects and planners can create the spaces to encourage new forms of social interaction and commerce. Architects become central to the question of how do we want to live both now and in the future. Alastair Fuad-luke[3] has argued for a holistic view of sustainability using well- being to link the environment to people, economics and the future. This makes architecture a privilege, and a creative responsibility. The following examples highlight how architects have risen to the challenge in stimulating ways both in reducing impacts and enhancing our relationship with buildings and each other. Julian Lindley

1. Ann Thorpe, *Architecture and Design versus Consumerism*, Earthscan, 2012.
2. Cameron Sinclair, "Sustainability as a matter of Survival: 10 Years of Architecture for Humanity", lecture at RSA, London.
3. A. Fuad-Luke, *Design Activism*, Earthscan, 2009.

能源构建未来

2050年。我们无法逃避全球变暖、温室气体排放、人口增长以及热岛效应对地球未来产生的重大影响。而仅仅两摄氏度之差，人类就可以远离难以忍受的生活条件。那么能源效率真的能拯救世界么？在本文中，这一可能的未来场景经过了科学分析和技术途径的审视。本期呈现的建造于世界各地的六个项目，在环境技术、能源节约和可持续性方面都给予了特殊的关注。

2050. No way out to escape. Global warming, GHG emissions, human growth and Heat Island Effect influence the future of the globe in a dramatic matter. Only two Celsius degrees divide the mankind from an unbearable living conditions. Can the energy efficiency save the world? In this article this possible future scenario is examined between scientific analyses and technical approaches. Six projects, built all around the world, are presented with special focus on the environmental technology, energy conservation and sustainability aspects.

能源效率：展望建筑未来
Energy Efficiency: Building for the Future

可持续性不是牺牲或减少的代名词。它反而是高设计质量的指标和对给定地理区域或国家潜在建筑市场的一种认可。在未来的场景中，至2050年全球气温至少上升2°C，而对高温湿热气候的预计似乎也是一种相当合理的假设。位于南极洲西部的伯德站，最近也给出自1958年以来年平均气温上升2.4°C的报道：是预期年平均上升值0.8°C[1]的三倍。因此，由全球174个国家签署的《京都议定书》中提到的减少温室气体排放[2]、增加能源效率以及可再生能源的使用俨然成为我们实现可持续发展未来的关键条件。燃烧非可再生燃料（如煤炭和石油）来产生电力或作为运输燃料一直是有害且有毒温室气体排放的原因所在。在新西兰，有记录显示该国年发电量为67 000BTU，其中的35%~40%都应用于建筑方面（参见mfe.gov.nz）。特别是商业建筑，因屋顶、墙壁、窗户、地基和漏气量是影响供热、制冷和换气负荷的主要元素，因此这类建筑耗费多达其能源的60%，用以保持适宜居住的最佳室内温度。也正因为如此，LEED议定书对可持续型建筑的围护结构设计的可信权值分配比例高达39%。这种关注被一个事实放大，即直至2050年，预计全球人口将增至90亿，比目前人口增加20多亿的事实。因此，为了能够满足更多人对现有能源的新需求，我们越来越需要重新考虑城市本身这一概念。因为

Sustainability is not a synonym for sacrifice or decrease. It is instead an indicator of high design quality and a recognition of the potential constructive market in a given geographical area or country. In the scenario of a future rise in global temperatures of at least 2°C by 2050, the prospect of a hot and humid climate seems a quite plausible hypothesis. The Byrd Station, in west Antarctica, has recently reported an increase in annual average temperature of 2.4 °C since 1958: three times higher than the expected average of 0.8°C[1]. Reduction of GHG (greenhouse gas) emissions[2], as reported in the *Kyoto Protocol* signed by 174 countries worldwide, and increased energy efficiency and the use of renewable energy resources have thus become crucial conditions for our future. The burning of non-renewable fuels such as coal and petroleum to generate electricity or as fuel for transportation has long been responsible for the release of harmful and toxic greenhouse gases. In New Zealand, it has been documented that 35%~40% of the country's 67.000 BTU of annual energy generation is utilized by buildings (cf. mfe.gov.nz). Specifically, commercial buildings spend up to 60% of their energy to maintain optimum indoor habitable temperatures, with roofs, walls, windows, foundations and air leakage being the primary elements affecting heating, cooling and ventilation loads. Precisely for this reason, the LEED Protocol has allotted up to 39% of its credit weight to the design of sustainable building envelopes. Such concerns are magnified

the future

一方面，冰川融化正导致海平面上升，而另一方面荒漠化的增加都将减少对人类活动（包括最值得关注的农业活动）十分重要的可居住土地的面积，尤其是当农业用地需要由38%增加至49%以维持世界人口增长的时候。环境技术、能源节约和可持续性都是未来可持续性设计工作日程的重要组成部分。接下来是对六个设计案例研究的分析，在这些案例中，上文提及的能源相关问题成就了项目自身的特色。

由Tectoniques建筑师事务所设计的Aquaterra环境中心，再现了欧洲最大的焦炭生产工厂（建于20世纪二十年代并于2002年关闭，占地953.0m²）之一的厂址重建之后的面貌。建筑师与Ilex景观建筑师事务所合作，在与先前工业活动无任何物理联系的情况下保留并还原了当地的集体回忆和遗产。该建筑已然成为公民与环境，以及工业历史与可持续未来之间关系的关键点。在环境问题和鼓励更多可持续性实践的普及和广泛应用，包括更好的废物处理方式、更具有可持续性的建筑设计以及从废物中获取更多的能源方面，Aquaterra环境中心可以说是一个具有启发作用并能唤起当地人觉悟的中心，而这些体现在小学生身上尤为明显。该中心将绝大部分的信息性内容都投放在气候干扰上。建筑的绿色屋顶将雨水转移至储水池以满足生活用水需求，包括温室灌溉，同时为光伏和风力发电系统提供一个平台。日光可以照进所有的室内房间中，同时巧妙的遮阳设计又能避免获得过度的热量。该中心已通过Minergie认证，该认证是瑞士注册的新型或翻新低能耗建筑的品质标签，类似于德国的KfW40（新型建筑）和KfW60（翻新建筑）标准。

由HHS Planer＋Architekten AG设计的能源地堡由曾经的纳粹独裁者始建于二战时期（1943年），这一曾经的防空高射炮掩体如今作为国际建筑展览（IBA）的一部分被保留了下来。2012年，该建筑经改造，成为一座为附近3000户家庭服务的可再生能源发电厂。该设施利用一个巨大的热水箱为Reiherstieg区提供气候友好型热能，同时将可再生能源输送至汉堡配电网。合理的结合太阳能（在屋顶和南面安装近3000m²的太阳能板）、沼气、木屑燃烧装置和附近工业工厂的废热产生85%的可再生电力，然后将其储存在地堡内，并输送至电网中。该项目最具创新性的特点在于大型的缓冲能源储存室，其总容量可达2百万公升（2000m³）。未来，德国北部多余的风能可能会被转变成热能储存在库中，或者在低风速的时候另一组合设备产生的热能会给予补足，以确保时时都有绿色电力的输出。

伦佐·皮亚诺建筑工作室设计的特兰托科学博物馆位于一个占地

by the fact that by 2050 the world population is expected to rise to about nine billion, no fewer than two billion more than the current population. Therefore, it will be increasingly necessary to rethink the concept of the city itself, in order to be able to match available resources to the new needs of more people. Glacier melt resulting in rising sea levels on the one hand and increased desertification on the other will subtract a significant expanse of inhabitable land for human activities, including, most concerningly, agriculture, even as agricultural land needs to increase from 38% to 49% to sustain the growing world population. Environmental technology, energy conservation, and sustainability are all crucial parts of any agenda for future sustainable design. What follows is an examination of six case studies in which the energy-related concerns mentioned above drive project characteristics.

Aquaterra Environmental Center by Tectoniques Architects rehabilitates one of the largest coke production plants in Europe (953.0 sqm), founded in the 1920s and shuttered in 2002. The architects, in collaboration with Ilex Landscape Architects, have preserved and restored local collective memory and heritage with no physical links to former industrial activities. The building has become a key point in the relationship between citizens and environment, and between the industrial past and a sustainable future. Aquaterra is a center for enlightening and raising the consciousness of locals, particularly schoolchildren, with regard to environmental questions and for encouraging the popularization and widespread application of more sustainable practices, including better waste handling, more sustainable architecture, and increased harvesting of energy from waste. A large part of the Center's informational content is devoted to climate disturbance. The building's green roof diverts rain into cisterns to serve non-potable water needs, including greenhouse irrigation, while providing a platform for photovoltaic and wind energy generation systems. Daylight reaches all interior rooms while strategic shading prevents excessive heat gain. The center has received Minergie Certification, a Swiss registered quality label for new and refurbished low-energy-consumption buildings, comparable to the German KfW40(new construction) and KfW60(refurbishment) standards.

Energy Bunker by HHS Planer＋Architekten AG was built during World War II by the Nazi dictatorship(1943), and this former anti-aircraft flak bunker has now been preserved as part of the International Building Exhibition(IBA). In 2012, it was converted into a renewable energy power plant serving 3000 nearby households. The facility uses an enormous heat tank to supply the Reiherstieg District with climate-friendly heat, while feeding renewable resources into the Hamburg Distribution's grid. Smart use of a combination of solar power(approximately 3000 sqm of solar panels on the roof and south face), biogas, a wood-chip burning unit

11公顷的棕色地带工业区之上，该场地先前是一个轮胎工厂。该博物馆项目属于一个更大的、竣工于2012年的、名为Le Albere的再开发项目的一部分。这个再开发项目提供居住、娱乐、社交和办公空间，同时发挥与市中心之间的纽带的作用。博物馆展品与阿尔卑斯山的生态系统和意大利北部的沿海环境紧密联系。该馆整体设计首要考虑的是能源效率。因此，该建筑的诸多立面都镶嵌绿色石头以吸收太阳能。用心设计的窗户和立面是为了优化其热性能，并且利用太阳能来减少照明设备的能量负荷。建筑的一些材料源于当地。屋顶上占地340m²的光伏电池板为建筑提供一些电力，地热系统提供节能的制冷和供热，而三联产电站既为该建筑，同时也为整个地区提供电力。通过为室内花园提供灌溉，雨水收集减少了建筑高达50%的用水量。而有限的停车位也鼓励人们使用公共交通工具。值得一提的是该节能建筑已经获得了LEED金奖认证。

由Szyszkowitz-Kowalski＋Partner ZT GmbH设计的格拉茨的斯巴克埃森－霍费尔储蓄银行竣工于2011年。这些庭院是对格拉茨典型室内庭院历史主题的重新诠释。玻璃板条组成的圆锥形场地影射了室内庭院的金字塔外形，并且借助于空中花园来防止夏季室内产生过高的温度。替代能源概念的特点在于利用地下水温的两口深井，它们结合安装于每层顶棚的热能反应系统，与带有快速反应的天花板表面一起，在冬季提供热能，并在夏季帮助散热。为此，建筑师并没有设计吊顶顶棚；反而所有必要的技术设施和隔音设备都安装于错误的楼层中，以实现高度的灵活性。此外，该建筑由一些带有短房间进深的狭窄结构组成，以确保该综合体建筑拥有良好的采光、通风和透明度。该建筑的设计还特别注重高效的降噪效果。

哥伦比亚林荫大道废水处理厂建于1950年，是一个处理波特兰组合废水和雨水的工业现场。受到景观以及该场地工业历史的启发，建筑师负责为工厂的工程师和公共接待区设计新的办公设施。Skylab建筑事务所设计的哥伦比亚大厦配备一个光伏发电系统和提供热能的现场热电联产发电厂，利用工厂净化水的机械系统起到了热泵的作用。在建筑外部，不锈钢遮阳物和天窗能够调整日光的照射量。同时，折叠屋顶和百叶窗也促进了日光照射和自然通风，并且可以对装有百叶窗的北向玻璃立面实施操作以在上班时间提供阳光和空气，并在夜晚帮助冷空气排出。雨水过滤和植被屋顶是环境服务局倡导的两个主要项目。波特兰市

and waste heat from nearby industry produces 85% renewable power, then to be stored in the bunker and fed into the electricity grid. The project's most innovative feature is its large-scale buffer energy storage chamber, with its total capacity of 2 million liters (2000 cubic meters). In the future, excess wind power from northern Germany could be transformed into heat in the tank, or heat from another combined unit could be fed at times of low wind, to ensure green electricity output at all times.

Trento Science Museum by Renzo Piano Building Workshop, located on an 11-hectare brownfield industrial area, was formerly a tire plant. The Museum is part of a much larger redevelopment project, completed in 2012 and called "Le Albere". This redevelopment provides space for housing, leisure, commerce and offices while serving as a link to the city center. The museum's exhibits are specifically tied to ecosystems of the Alpine and coastal environments of northern Italy. Energy efficiency was a top priority in the Museum's overall design. Many of the facility's facades are clad in green stone to absorb solar energy. The windows and facades are carefully designed to optimize thermal properties, and daylight is harnessed to reduce the energy load from lighting. Some materials were sourced locally. The roof houses 340 sqm of photovoltaic panels that produce some of the building's electricity, a geothermal system provides energy-efficient cooling and heating, and a tri-generation power plant serves both the building and the entire district. Rainwater collection reduces the building's water usage by up to 50% by supplying irrigation for the indoor garden. Parking is limited to encourage use of public transportation. This energy efficient building has obtained LEED Gold certification.

Sparkassenhöfe Graz by Szyszkowitz-Kowalski＋Partner ZT GmbH was completed in 2011. These courtyards are a re-interpretation of the historical theme of the interior courtyards typical to Graz. A conic field of glass slats alludes to the pyramid form of the interior courtyard and provides protection from overheating during the summer with the help of hanging gardens. An alternative energy concept features two deep wells that take advantage of groundwater temperature, helping a thermo-active system installed in each of the storey's ceilings to provide heating in winter and cooling in summer, with additional quick-reaction surfaces on the ceilings. For this reason there are no suspended ceilings; instead, all necessary technical installations and soundproofing facilities are installed in false floors that have a high level of flexibility. Additionally, the architecture consists of narrow structures with short room depths that in themselves ensure good lighting, ventilation and transparency throughout the building complex. Special attention is given to highly effective noise reduction.

The Columbia Boulevard Wastewater Treatment Plant was constructed in 1950 as an industrial site for treating Portland's com-

在城市规划、公共事业和建筑方面推广可持续性实践是具有一定历史性的，其市政绿色建筑政策要求公共建筑体现可持续性实线并且至少要通过LEED金奖认证。

由米库设计工作室开发的利用太阳能电池板的零能耗学校竣工于2013年，该港口学校的设计目的在于实现零能源消耗，以成为该地区可持续发展的标志性建筑。南向的教室和操场优化利用了被动式太阳能，并且增加了适合安装太阳能板的面积，使其充分融入到带顶操场区域的建筑当中。而封闭内部庭院下方的诸多教学区域阳光充足，并且环境十分安静。

对这六个案例的审视使我们弄清楚了建筑尺度之上城市设计的共同偏好，以便提高我们对当前和未来现有资源需求影响的认知。这种紧张感无疑在激进且弹性的策略成果方面产生了丰富的创造性以及形式和技术上的创新，而它们在概念上就有别于19世纪的花园城市和城市扩张策略。结论似乎很清晰，只有一个具有高密度和低辐射的城市模型才是尤纳·弗里德曼众多假设的唯一答案。因为他说过，"想象整个欧洲集中于100或200座城市，整个中国集中于200座城市以及整个世界集中于1000多座城市并不是错误的想法"[3]。目前，我们极需增加城市住房的密度并实现能源的自给自足。一个明确保证至少实现短期效果的应急策略是通过"能源加"软件的介入来改进市中心建筑的构造的，以此将市区转变成发电装置[4]。

bined wastewater and stormwater. Inspired by the landscape and the site's industrial past, the architect was tasked with creating new office facilities for the plant's engineers and public reception areas. The Columbia Building by Skylab Architecture has a photovoltaic system and an on-site co-generation plant for heating. The mechanical system works as a heat pump, utilizing the plant's processed water. Outside, stainless steel shades and clerestory windows modulate daylight. Daylight and natural ventilation are also facilitated by a folded roof and a north-facing louvered glass facade which may be manipulated to supply light and air during working hours and to facilitate flush cooling at night. Storm water filtration and vegetated roofs are two major programs prompted by the Bureau of Environmental Services. The City of Portland has historically promoted sustainable practices in city planning, utilities, and architecture, and its municipal Green Building Policy requires public buildings to incorporate sustainable design practices to achieve a minimum of LEED Gold certification.

Zero Energy School in Solar Panels by Mikou Design Studio was completed in 2013, and the docks schools have been created to consume zero energy in order to become an emblem of the area's sustainable development. The southern orientation of classrooms and playgrounds makes optimal use of passive solar energy and increases the amount of surface suitable for the photovoltaic panels integrated into the architecture of the covered playground areas. The various teaching areas, below sheltered internal patios, are open to light and quiet. The examination of these six cases makes clear a common preference for the urban over the architectural scale so as to increase awareness of the impact of current and future needs with respect to currently available resources. This tension has no doubt generated much creativity as well as formal and technological innovation in the production of radical, resilient strategies, conceptually distinct from the nineteenth-century garden city and urban sprawl. The conclusions seem to bring into clarity that a city model marked by high density and low emissivity is the only possible answer among those postulated by Yona Friedman. He argues, "It is not wrong to imagine the whole of Europe concentrated in 100 or 120 cities, 200 in the whole of China and the whole world grouped around a thousand"[3]. Increasing urban housing density and energy self-sufficiency are now absolute needs. One emergent strategy that clearly guarantees at least short-term results is to retrofit the fabric of downtown buildings through "Energy-Plus" interventions, thus transforming urban districts into power generators[4]. Fabrizio Aimar

1. Greene, C.H., "The Winters of Our Discontent"[On-line], Scientific American 307: 50-55 http://dx.doi.org/10.1038/scientificamerican1212-50, 2012.
2. Kyoto Protocol, "United Nations Framework Convention on Climate Change" [On-line], https://unfccc.int/kyoto_protocol/items/2830.php.
3. Yona Friedman, *Utopies Réalisables*, Edition l'èclat, Paris, France, 2003.
4. Dr. Alessandro Melis, Pengfei Li, Shivani Khanna. *Energy-Plus Downtown*, National Institute of Creative Arts and Industries(N.I.C.A.I.), University of Auckland, School of Planning and Architecture, 2014.

Aquaterra环境中心

Tectoniques Architects

在当地征集了民意之后，海宁－卡尔万联合城市管理局采用了"Aquaterra"来为这座环境保护与可持续发展中心命名。该设施位于一个大型公园内，该场所原来是德罗库尔焦煤工厂。工厂建于1925年，是欧洲最大的焦煤生产工厂之一。在20世纪末煤炭和钢铁工业变革之后，工厂最终在2002年关闭。

所有的建筑都被迅速拆毁了，只留下一个大平台和一些矿渣堆。被沾染了的地面和矿渣堆、丘陵和山脉，构成了毫无生机却极具标志性的地貌特征，远远看去，却成为了这块平地上唯一的慰藉。

当今，建筑师和景观设计师有义务维护和复原这个地区共有的回忆和历史遗产，同时还要避免与之前的工业活动有任何实际联系。为了免于成为一座博物馆，这片区域必须找到新的出路。当地居民和这片土地的关系已不再融洽，而生态化则能够使之重修于好。建造这座资源建筑的目的是为了引起参观者对环境问题的关注。建筑师将这个项目当作一项实证方案和实物实验来设计。经过了Minergie认证，在其木砖外层及透镜结构之下，这座建筑包含了足以确保房屋功能良好运转和温暖舒适的所有能源系统。

Aquaterra中心是一个更大的公园项目的一部分：由Ilex景观建筑师事务所设计的群岛公园。在这片45公顷的空地上，方案提议建造一组群岛和人工湖，并由构成这片区域的主要路径连接。为创造一处好玩的、隐秘的自然环境，公园仍保留了地面被沾染的痕迹，别具一格地将想象与现实联系到一起。

Aquaterra中心象征了蜕变的第二个阶段——"进入公园"，包括建筑及其伴随的景观。其战略性位置使这座建筑在时代和地区的发展中占有举足轻重的地位。其透镜结构与整体设计和谐辉映。

坐落于群岛公园的西北部，这座建筑成为周围多种城市规划、正在进行的道路和景观项目的中心。按照欧式方案，城市居民区毗邻由矿区具有代表性的小型住宅集群组成的居民区。道路的连通方案已包含在纵贯南北的通道升级项目中，是不可缺少的联系。最后，"进入公园"景观美化方案增加了矿堆渣内栽种植被的类别。

这座建筑是人与自然之间、工业的历史和未来之间、以及城镇和景观之间的一个关系定位点。

Aquaterra中心作为一项教学资源，唤起当地居民，尤其是小学生，关注各种环境问题，倡导类似垃圾分类、可持续性住宅和消耗内含能源这类良好实践的推广和普及。温暖的矿渣堆土质加速了外来植物和种子的茂盛生长，由此可见，设计者在抵御气候干扰方面投入了很大一部分精力。黑色的土壤与外来植物形成反差，在创造了一种非凡的视觉效果的同时，也提供了天然的污染防治系统。

Aquaterra Environmental Center

After a call for ideas addressed to the local residents, the Center for the Environment & Sustainable Development of the Hénin-Carvin Joint Urban Authority adopted the name "Aquaterra". The facility is in the middle of a large park on the site of the former Drocourt Coking Plant. Founded in 1925, this was one of Europe's largest coke production plants. After changes in the coal and steel industries in the late 20th century, the plant finally closed in 2002.
All of the developments were quickly demolished, leaving just a large platform and some terrils (slag heaps). The marked, polluted ground and the slag heaps, mountains and hills, composed of inert but emblematic masses in this flat landscape, are the only relief visible from a long distance.
Today, architects and landscape designers have a duty to preserve and restore local collective memory and heritage, but without

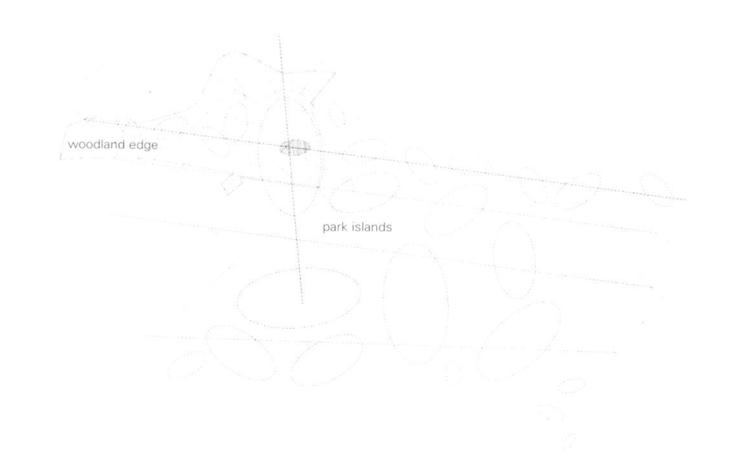

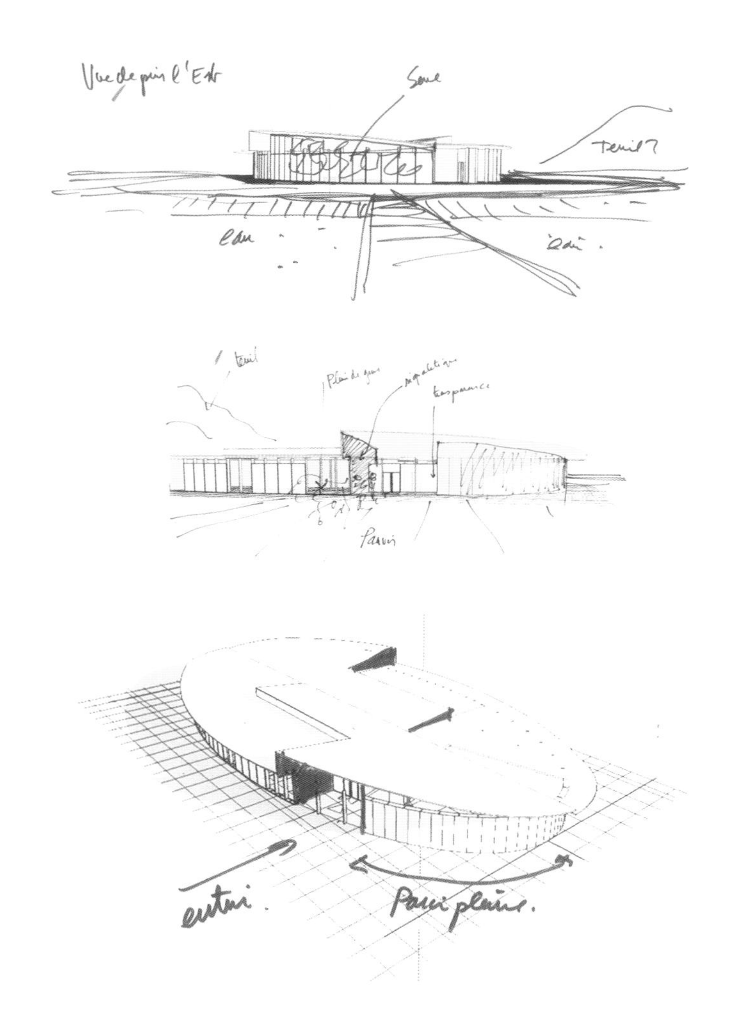

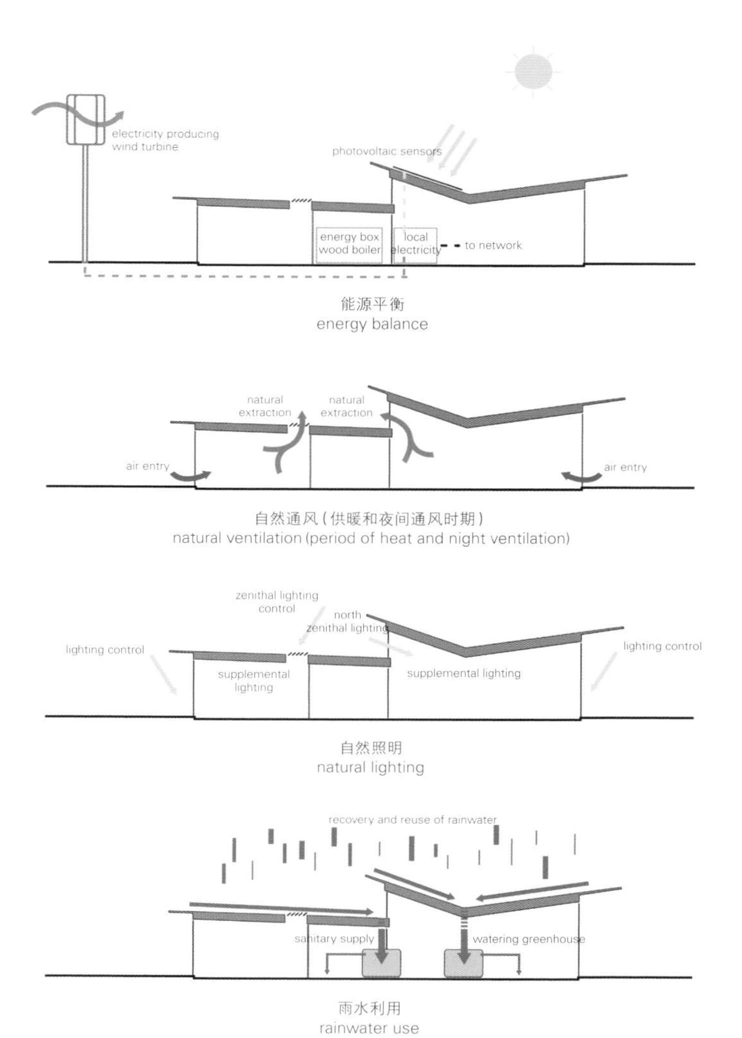

能源平衡
energy balance

自然通风（供暖和夜间通风时期）
natural ventilation (period of heat and night ventilation)

自然照明
natural lighting

雨水利用
rainwater use

having any physical links with the former industrial activities. To avoid becoming a museum, this area must find itself a new purpose. Ecology is used as a means to reconcile the local people and this place, which have fallen out of love with each other. The aim of this resource building is to make visitors aware of environmental questions. The architects have designed it as a demonstrational scheme and a life-size experiment. With Minergie certification, behind its original skin of wood bricks and its lens form, it contains all the energy systems that ensure good practices and thermal comfort.

Aquaterra is part of a much larger park project: Islands Park designed by the Ilex Landscape Architects. On the 45 hectare site, the scheme proposes a play of islands and artificial lakes, linked together by the major routes that structure the area. Creating playful, esoteric nature, off-beat in relation to the imagination and normal practices, the park reverses the image of a site that is very badly marked.

Aquaterra symbolizes the second phase of the metamorphosis called L'Orée du Parc, which includes the building and its accompanying landscape. It is strategically located, and the building occupies a pivotal position in the time and space of the development. Its lens' form is in harmony with the overall design.

In the north-west of the Islands Park, the building is central to various urban plannings, roads and landscaping projects currently in progress. Urbanité, according to Europan's scheme, is adjacent to neighborhoods made up of clusters of small houses that are typical of mining areas. Connexions, a scheme which consists in upgrading of the north-south access road, are the indispensable link. Lastly, the LOrée du Parc landscaping scheme adds to planting of the slag heaps.

The building is the anchor point for the relationship between people and nature, between the industrial history and the future, and between the town and the landscape.

Aquaterra is a resource for educating and raising the awareness of the local people, particularly schoolchildren, concerning all environmental questions and encouraging the popularization and widespread application of good practices, such as waste sorting, sustainable housing, and waste embodied energy. A very large part is devoted to climate disturbance, which is spectacularly illustrated by the blooming of exotic plants and seeds accelerated by the warm ground of the slag heaps. The contrast between black soil and exotic plants creates an extraordinary visual scene as well as provides natural pollution control.

东立面 east elevation

西立面 west elevation

Aquaterra

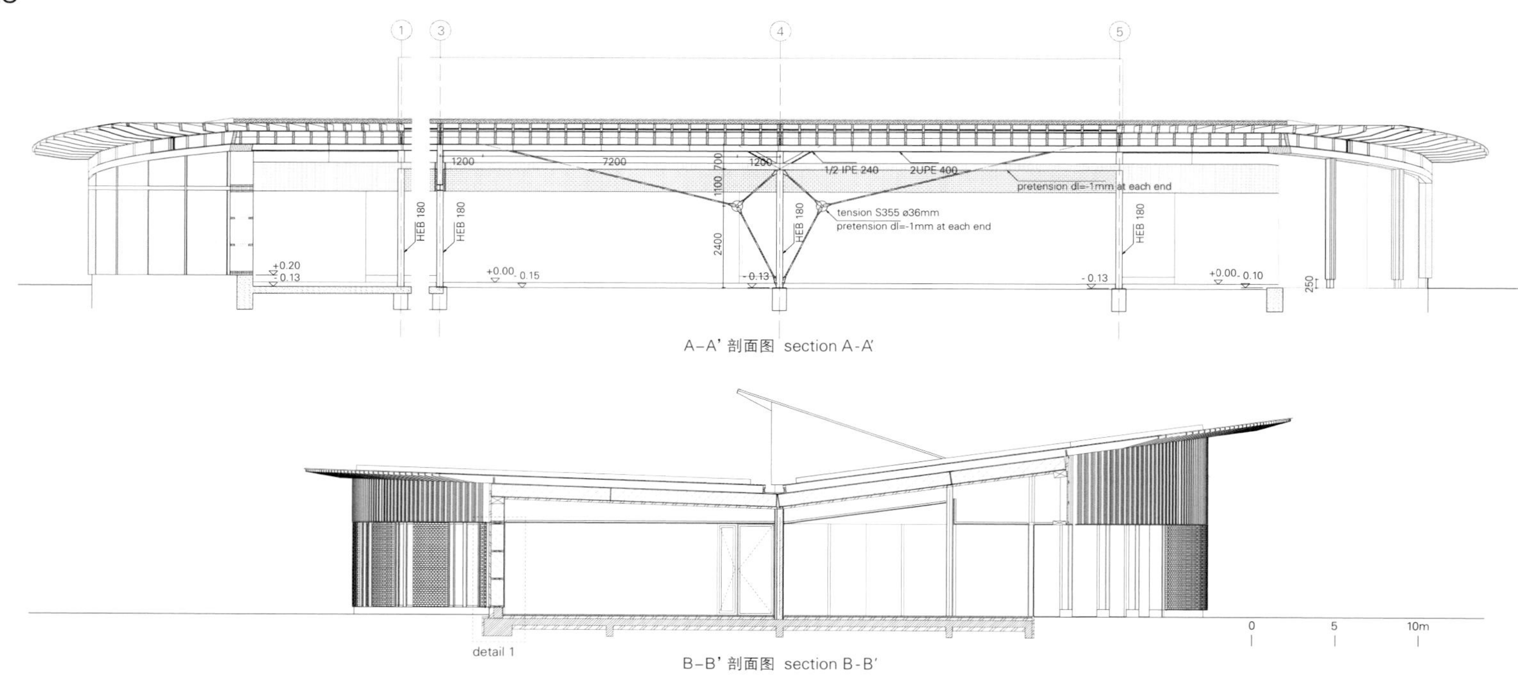

A–A' 剖面图 section A-A'

B–B' 剖面图 section B-B'

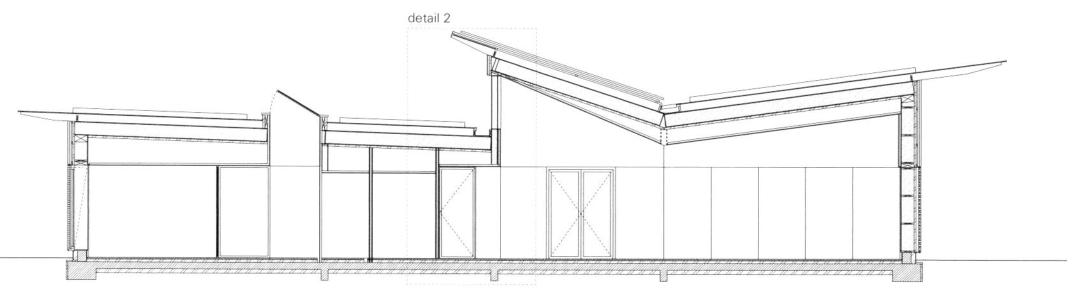

C–C' 剖面图 section C-C'

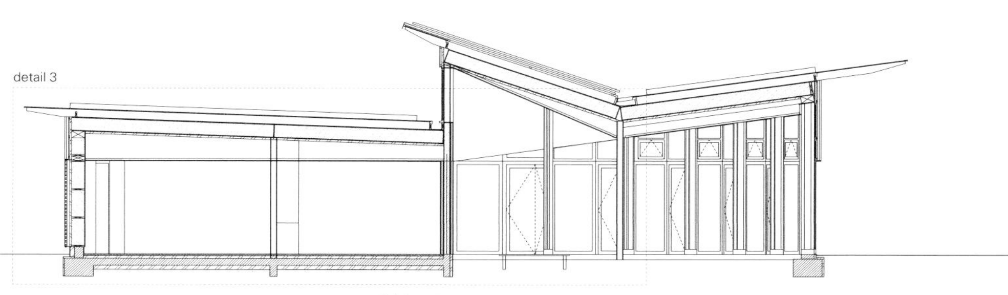

D–D' 剖面图 section D-D'

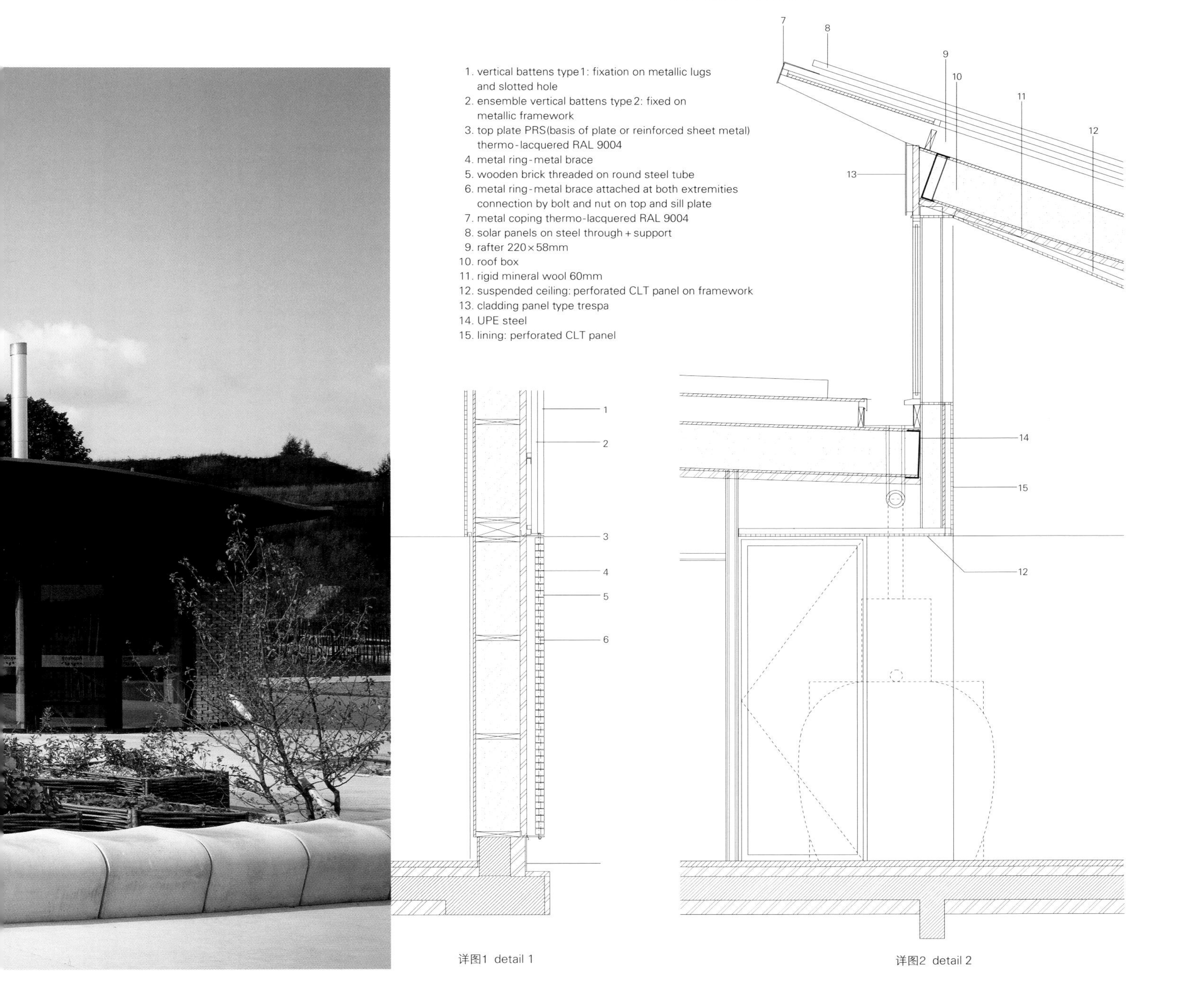

1. vertical battens type1: fixation on metallic lugs and slotted hole
2. ensemble vertical battens type2: fixed on metallic framework
3. top plate PRS(basis of plate or reinforced sheet metal) thermo-lacquered RAL 9004
4. metal ring-metal brace
5. wooden brick threaded on round steel tube
6. metal ring-metal brace attached at both extremities connection by bolt and nut on top and sill plate
7. metal coping thermo-lacquered RAL 9004
8. solar panels on steel through + support
9. rafter 220×58mm
10. roof box
11. rigid mineral wool 60mm
12. suspended ceiling: perforated CLT panel on framework
13. cladding panel type trespa
14. UPE steel
15. lining: perforated CLT panel

详图1 detail 1

详图2 detail 2

项目名称：Aquaterra environmental center
地点：Boulevard des Frères Leterme, 62110 Hénin -Beaumont, France
建筑师：Tectoniques Architects
土木工程：Balestra
木结构：Arborescence
景观建筑师：Ilex
流体和环境设计师：Indiggo
道路、设施以及室外作业：Maning
设计办公室（质量控制）：Veritas
场地监管&协作：Egis Bâtiment Management
健康与安保：Dekra / 暖通空调设计：MGC
甲方：Communauté d'Agglomération Hénin Carvin – CAHC
光伏设备：Forclum (Eiffage)
光伏产量：25 kW / 235m²
用地面积：953m² / 总建筑面积：576m²
设计时间：2011 / 竣工时间：2013
摄影师：©Julien Lanoo (courtesy of the architect)

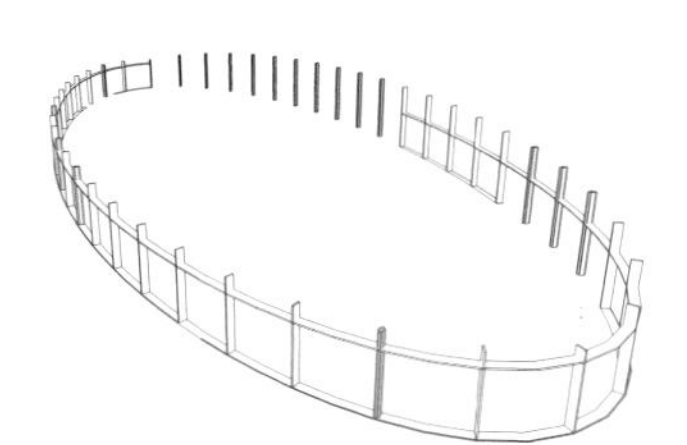

钢筋混凝土+木结构
RC + wood structure

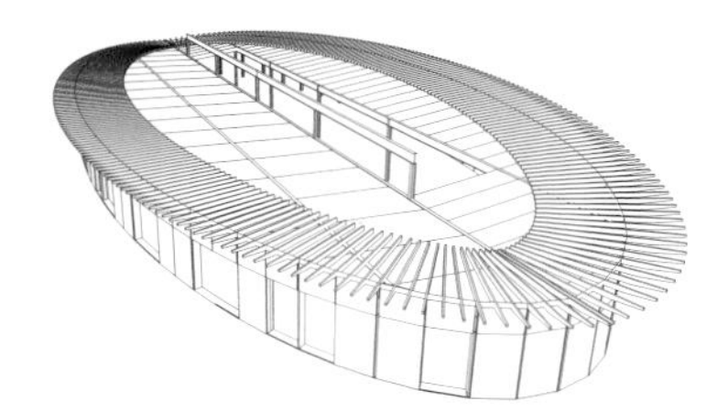

钢筋混凝土+木结构+金属构件+
水平嵌板+垂直嵌板+椽
RC + wood structure + metal
+ horizontal panel + vertical panel
+ rafter

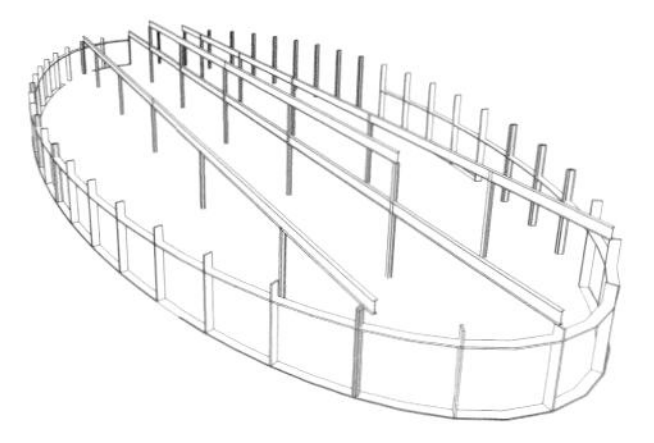

钢筋混凝土+木结构+金属构件
RC + wood structure + metal

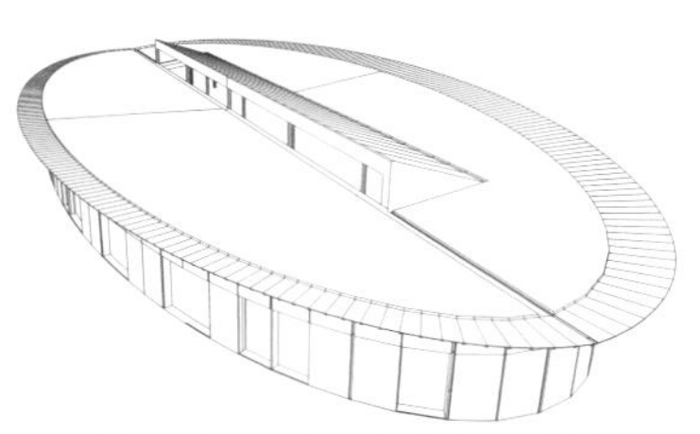

钢筋混凝土+木结构+金属构件+水平嵌板+
垂直嵌板+椽+屋顶
RC + wood structure + metal
+ horizontal panel + vertical panel
+ rafter + roof

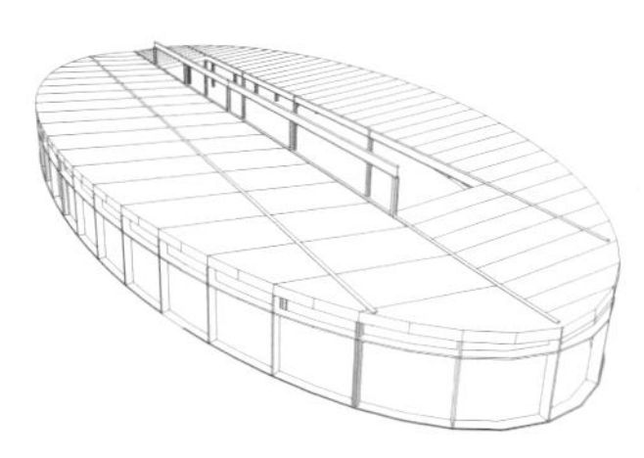

钢筋混凝土+木结构+金属构件+水平嵌板
RC + wood structure + metal
+ horizontal panel

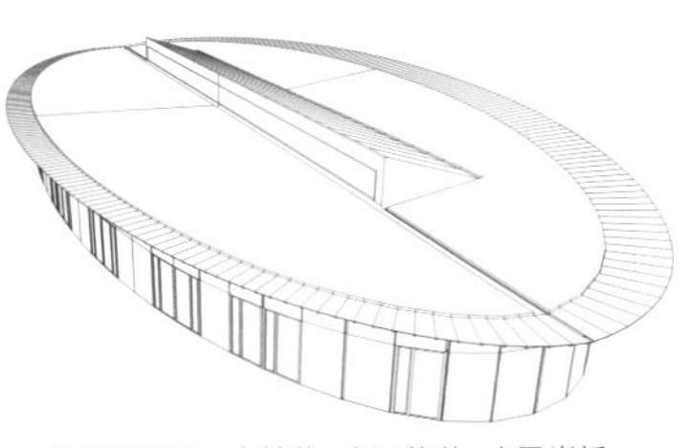

钢筋混凝土+木结构+金属构件+水平嵌板+
垂直嵌板+椽+屋顶+结合件
RC + wood structure + metal + horizontal panel
+ vertical panel + rafter + roof + joiner

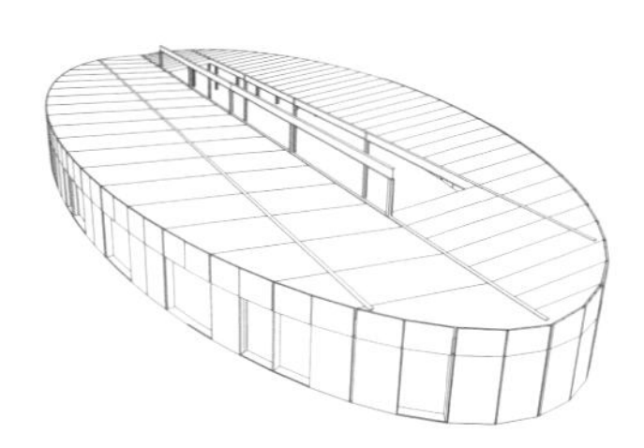

钢筋混凝土+木结构+金属构件+
水平嵌板+垂直嵌板
RC + wood structure + metal
+ horizontal panel + vertical panel

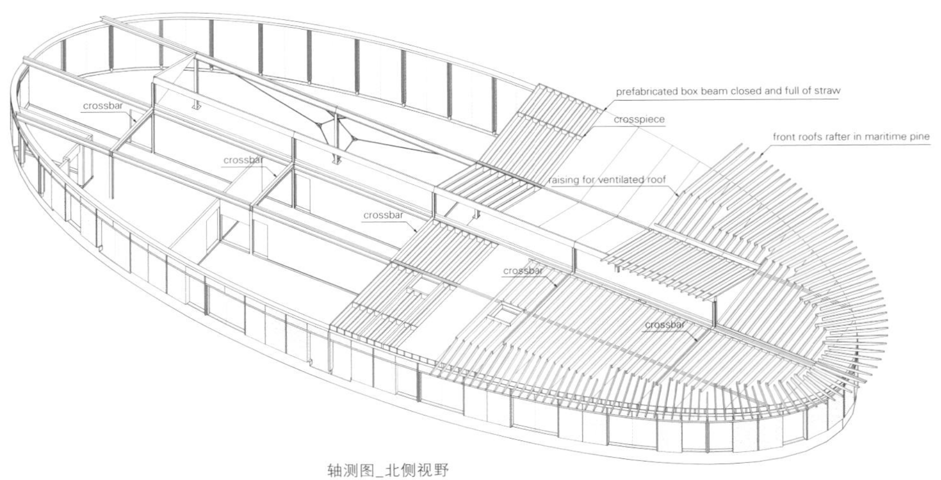

轴测图_北侧视野
axonometric _ north view

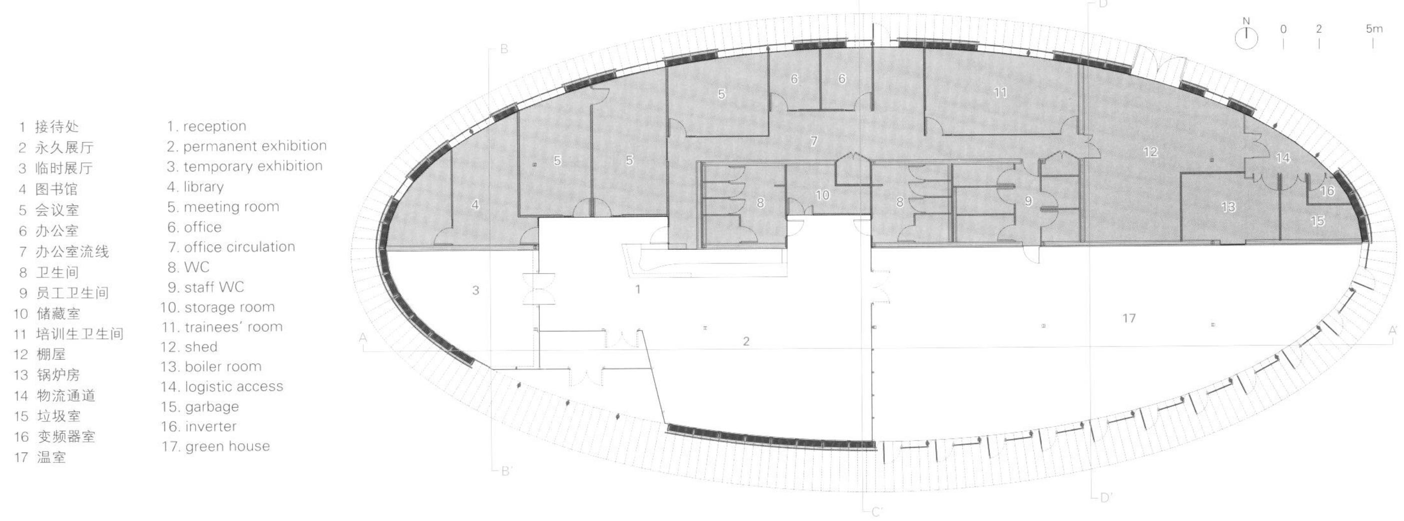

一层 first floor

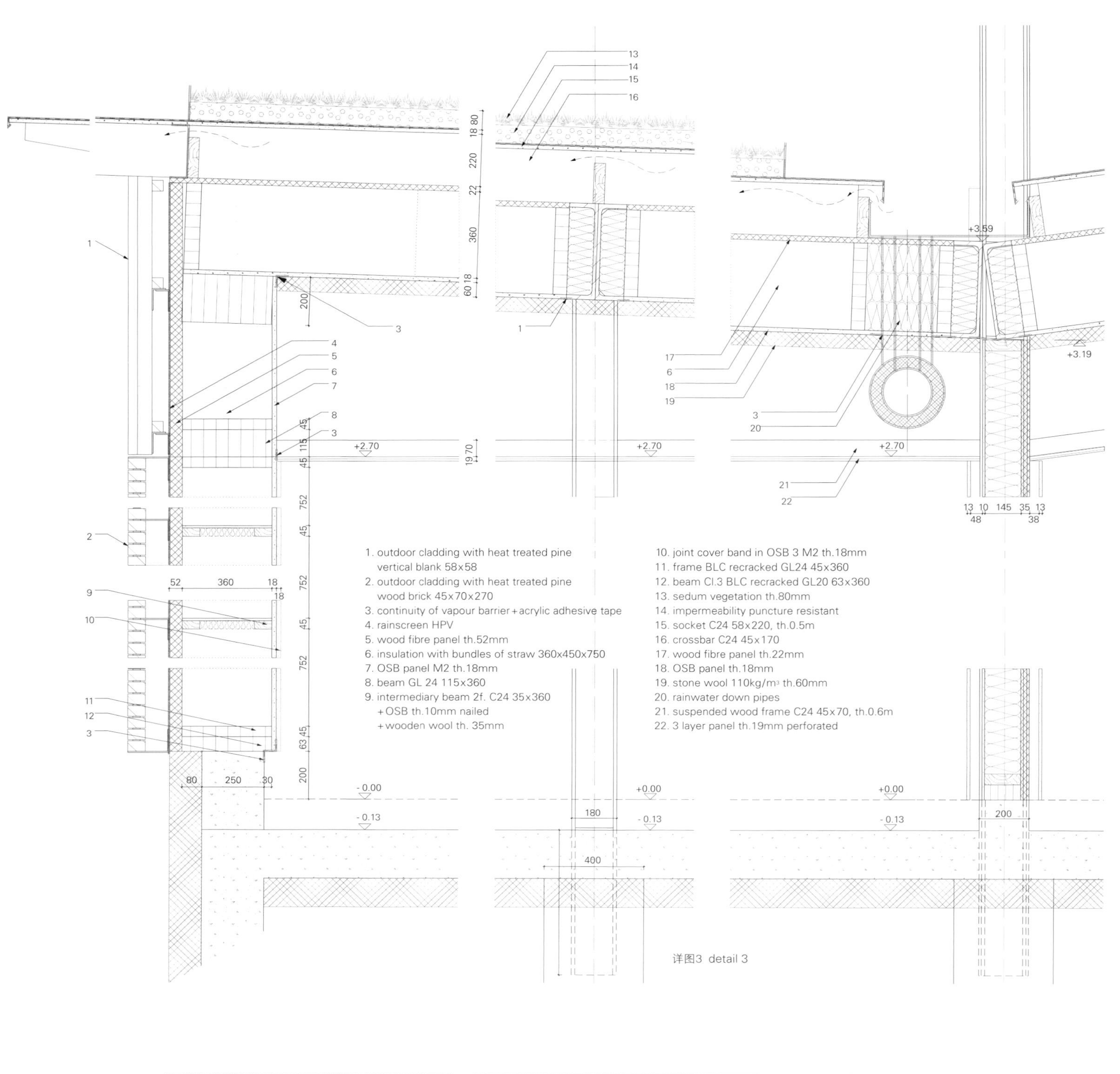

1. outdoor cladding with heat treated pine vertical blank 58x58
2. outdoor cladding with heat treated pine wood brick 45x70x270
3. continuity of vapour barrier+acrylic adhesive tape
4. rainscreen HPV
5. wood fibre panel th.52mm
6. insulation with bundles of straw 360x450x750
7. OSB panel M2 th.18mm
8. beam GL 24 115x360
9. intermediary beam 2f. C24 35x360 +OSB th.10mm nailed +wooden wool th. 35mm
10. joint cover band in OSB 3 M2 th.18mm
11. frame BLC recracked GL24 45x360
12. beam Cl.3 BLC recracked GL20 63x360
13. sedum vegetation th.80mm
14. impermeability puncture resistant
15. socket C24 58x220, th.0.5m
16. crossbar C24 45x170
17. wood fibre panel th.22mm
18. OSB panel th.18mm
19. stone wool 110kg/m3 th.60mm
20. rainwater down pipes
21. suspended wood frame C24 45x70, th.0.6m
22. 3 layer panel th.19mm perforated

详图3 detail 3

Cactus San Pedro

哥伦比亚某建筑

Skylab Architecture

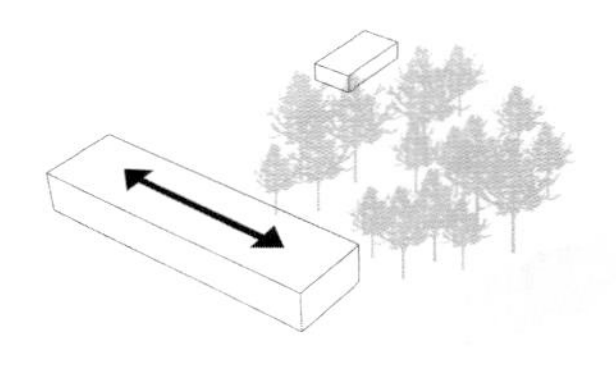

双压式办公室建筑
double loaded office

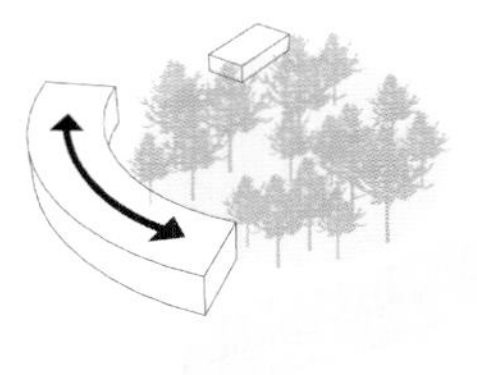

遵照大众风格
conform to commons

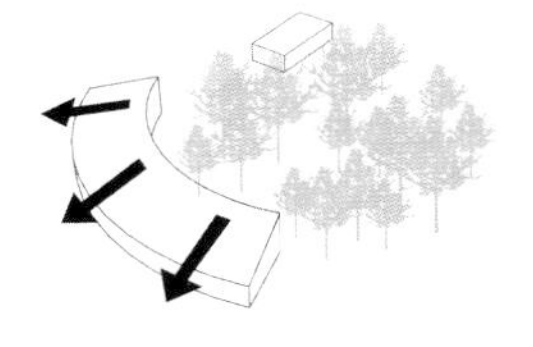

设置斜坡，以排放水流
slope roof to drain water

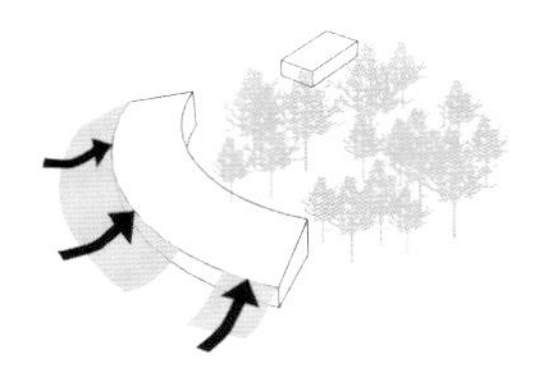

设置景观坡台，以阻止噪音，提供安全性
landscaped berm to block noise and create security

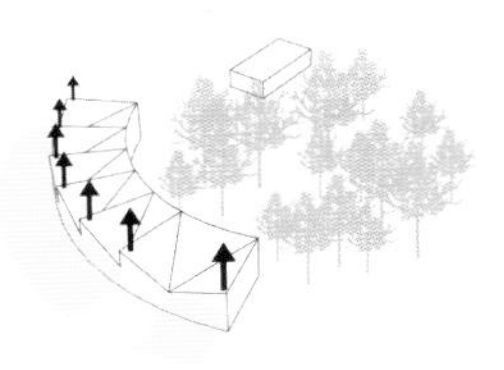

屋顶进行折叠，以促进照明和自然通风
roof folded to promote daylighting and natural ventilation

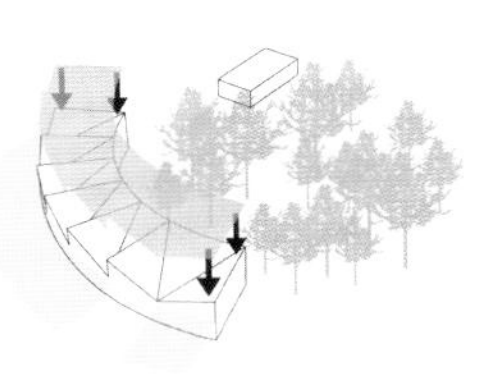

景观融入屋顶，以恢复自然环境
landscape integrated to restore natural habitat

哥伦比亚大道的废水处理厂建于1950年，作为一处工业场所，负责综合处理波特兰市的污水和雨水。近期，这项市政工程通过努力突出可持续发展的基础设施的重要性，来变得日益公开。设计建议为一个工程技术人员提供办公空间，他在过去的16年里都是在施工现场的移动拖车里工作的。该项目将为工厂塑造一个公众的形象，在为工程师和施工管理人员提供办公空间之余，还能大方地接待游客来访，承办公众会议。

单层的建筑的结构为现浇混凝土，生态屋顶和可视雨水收集装置是其突出的特点。这座建筑的雨水收集系统和标示性外观都在有意彰显着它与哥伦比亚沼泽生态系统及该地区流域更深层次的关系。

这座建筑是一座地貌、本土植物、严谨的几何图形及耐久性建筑体系的集合，其设计灵感来自于当地景观以及这里曾是工业旧址的现实条件。该场所以一个绿色空间为中心构筑而成，代替了原有的辐射式道路，通向工厂。此举提高了车辆的流通率、工厂的安全性和停车场布局的合理性。

建筑利用光伏发电系统，同时也利用现场的热电联产发电厂来获取电能。其机械系统是一个蒸汽泵系统，充分利用了工厂的工艺用水资源。

外部的不锈钢遮阳百叶窗和天窗系统形成了可调节的日光照明系统，与朝向北方的全玻璃外立面共同作用，来连接建筑内部与中央绿色空间。

历史上，波特兰市在城市规划、公共设施以及建筑方面已有过提倡可持续发展的实践。最近的绿色建筑政策要求公共建筑必须包含可持续发展的设计实践，至少要达到LEED金牌认证的最低标准。环境服务局需要确保流回到哥伦比亚沼泽地的雨水质量。环境服务局利用雨水过滤设施和植被覆盖的屋顶这两个主要项目，确保了波特兰地区在建设与发展过程中雨水的可持续发展管理。

The Columbia Building

The Columbia Boulevard Wastewater Treatment Plant was constructed in 1950 as an industrial site to treat the City of Portland's combined wastewater and stormwater. In recent times, this municipal project has become increasingly public through efforts to highlight the importance of sustainable infrastructure. The design proposed an office space for an engineering staff that over the past 16 years had worked in portable trailers at the site. It would also create a public face for the plant, programming generous visi-

项目名称：The Columbia Building
地点：Portland, Oregon, USA
建筑师：Skylab Architecture
工程师：Solarc Architecture and Engineering. Inc., Catena Consulting Engineers
照明工程师：Biella Lighting
景观建筑师：2-ink Studio Landscape Architecture
土木工程师：Vigil-Agrimis Inc.
环境平面设计：The Felt Hat
承包商：Skanska USA Building
甲方：The City of Portland Bureau of Environmental Services
用地面积：20,773m² 总建筑面积：1,067m²
有效楼层面积：1,394m²
设计时间：2010—2012 / 施工时间：2012—2013
摄影师：©Jeremy Bittermann(courtesy of the architect)

a.将游客停车场与非高峰时段停车场结合，以形成64 373m长的小径。
b.设有池塘观景平台。
c.建筑成为围墙线的一部分。

项目建成后的公共交界面
public interface after project
a. combine visitor parking with off-hour parking for forty mile path use.
b. allow for pond viewing platform.
c. building becomes part of fence line.

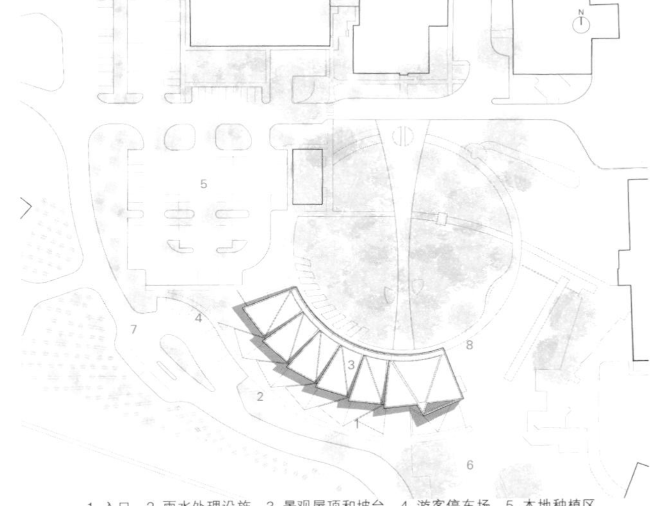

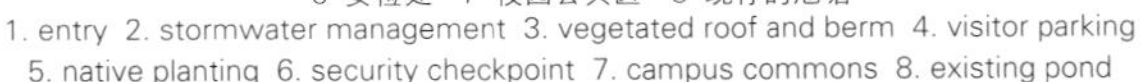

1 入口 2 雨水处理设施 3 景观屋顶和坡台 4 游客停车场 5 本地种植区 6 安检处 7 校园公共区 8 现存的池塘
1. entry 2. stormwater management 3. vegetated roof and berm 4. visitor parking 5. native planting 6. security checkpoint 7. campus commons 8. existing pond

a.植物成为本地景观的一部分。
b.对自然环境加以保护，并且进行了重建。
c.整修的花园成为中央园区公共区。
d.设置平台，为未来的扩建以及公共发展做准备。

项目建成之后的景观
landscape after project
a. plant becomes part of native landscape.
b. natural habitat is preserved and reestablished.
c. manicured garden is captured as central campus commons.
d. stage is set for future expansion and commons oriented development.

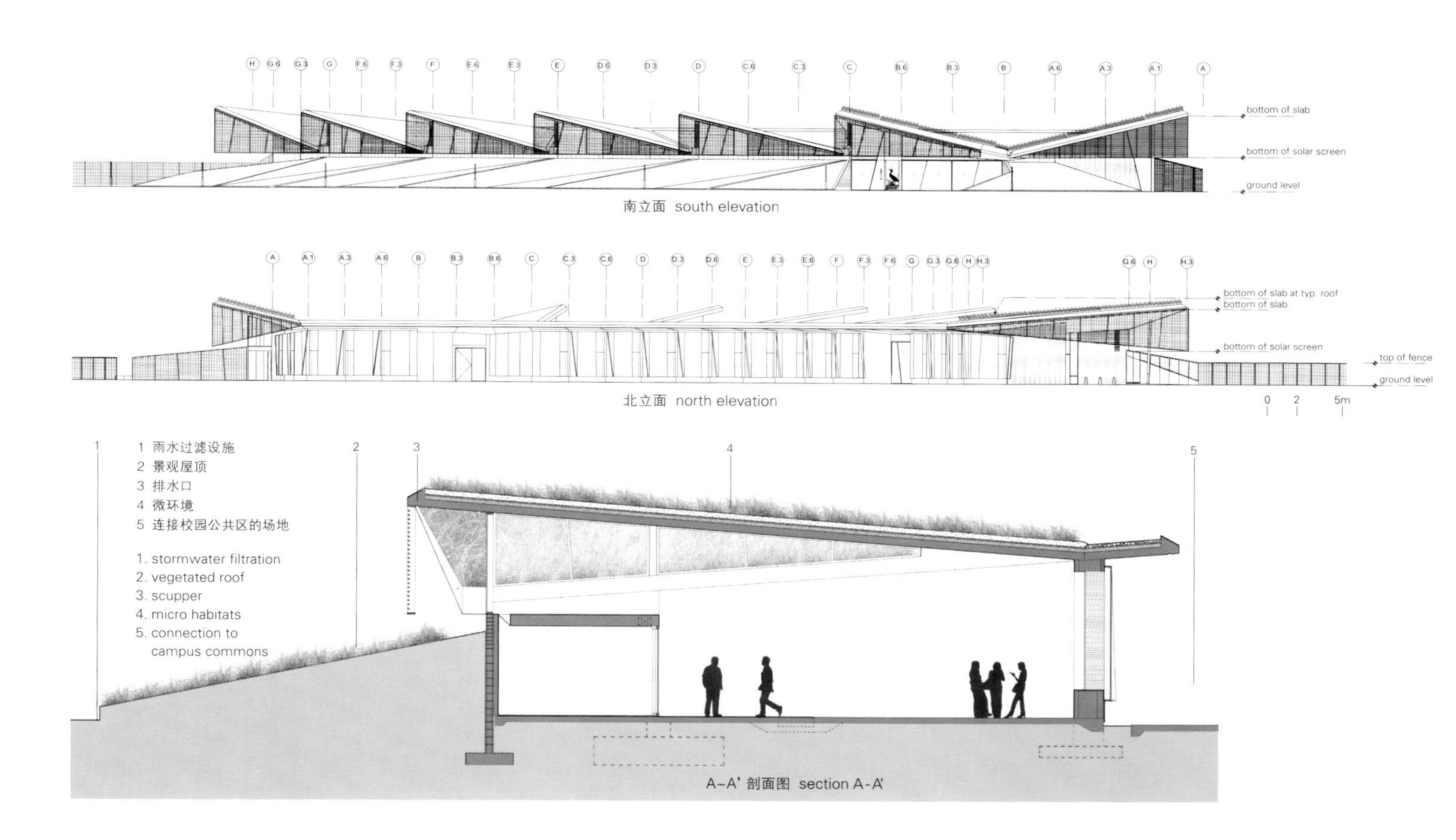
bottom of slab
bottom of solar screen
ground level
南立面 south elevation
bottom of slab at typ. roof
bottom of slab
bottom of solar screen
top of fence
ground level
北立面 north elevation
0 2 5m
1 雨水过滤设施
2 景观屋顶
3 排水口
4 微环境
5 连接校园公共区的场地
1. stormwater filtration
2. vegetated roof
3. scupper
4. micro habitats
5. connection to campus commons
A–A' 剖面图 section A-A'

tor reception and public meeting space along with office space for engineers and construction management staff.

The single story structure is cast-in-place concrete and features an eco-roof and a visible stormwater collection system. As an intentional demonstration, the building's stormwater system and signage combine to foster a deeper relationship with the Columbia Slough ecosystem and the regional watershed.

Inspired by the native landscape and the site's industrial past, the building is a combination of landform, indigenous planting, formal geometry, and durable construction systems. The site is organized around a central green space that replaces an original axial road leading to the plant, a move that has improved vehicular circulation, plant security, and parking organization.

The building has a photovoltaic system and also benefits from an on-site co-generation plant for power. The mechanical system is a heat pump system that utilizes the plant's process water source. Exterior stainless steel solar shades and a system of clearstory windows create modulated daylighting that work in combination with a fully glazed north facade to connect the interior spaces with the central green space.

The City of Portland has historically promoted sustainable practices in city planning, utilities, and architecture. The current Green Building Policy requires public buildings to incorporate sustainable design practices to achieve a minimum of LEED Gold Certification. The Bureau of Environmental Services protects the quality of stormwater returned to the Columbia Slough. Stormwater filtrating facilities and vegetated roofs are two major programs that the Bureau of Environmental Services utilizes to ensure sustainable stormwater management in Portland's construction and development.

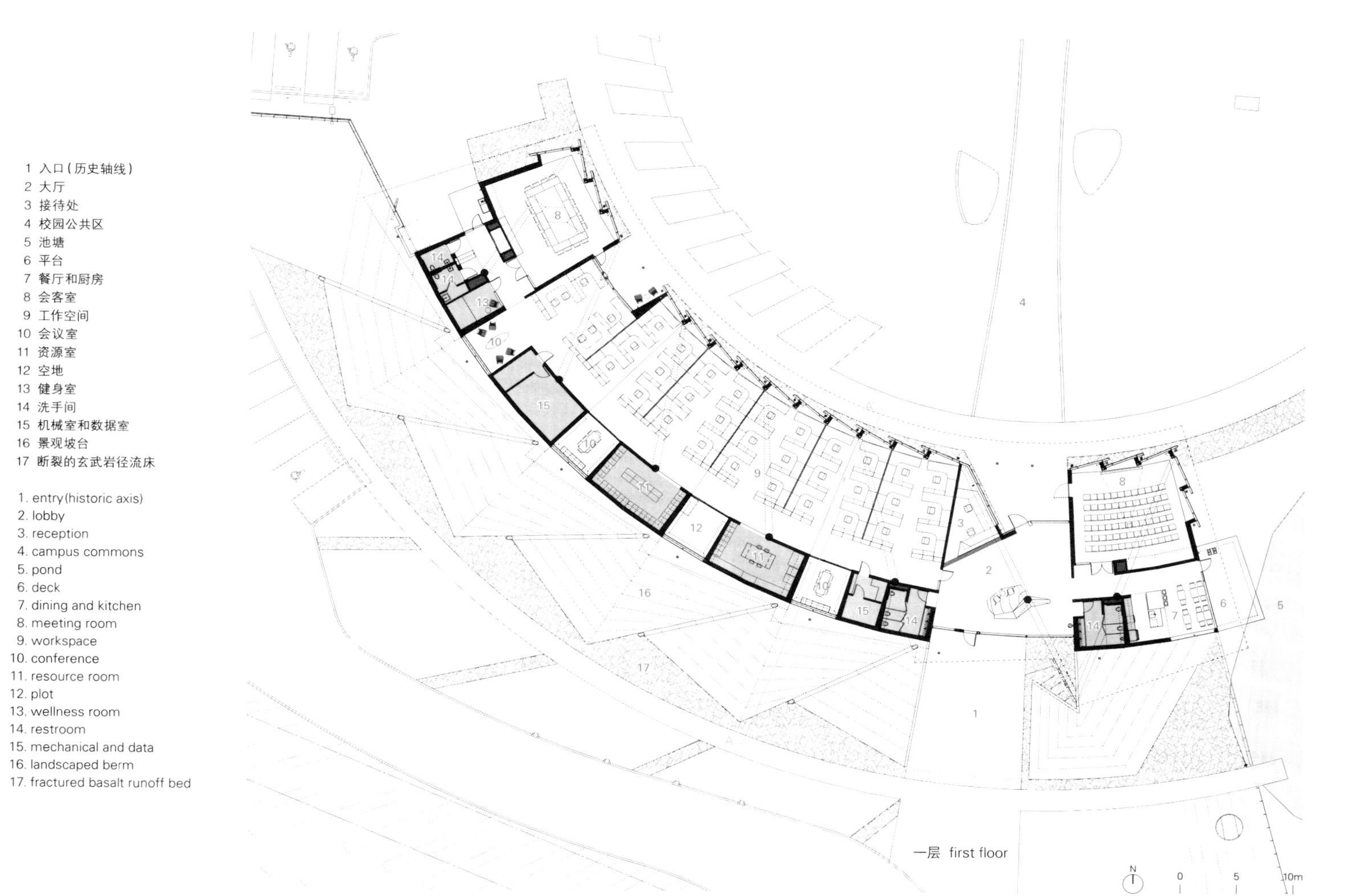

一层 first floor

能源地堡

HHS Planer+Architekten AG

威廉斯堡的原防空地堡经改造后成为"可再生威廉斯堡"气候保护理念的一个标志。如今，这座自二战结束后便几乎不再投入使用的历史遗迹在汉堡国际建筑展时期进行了翻修，被改造成一个使用可再生能源的发电站，并配备一个大型的储热器。该发电站为Reiherstieg区提供气候环保型热源，同时向汉堡配电网提供可再生能源电力。这座位于Neuhöfer街上的防空地堡建于1943年，用以展现战争中后方人民的英勇。成千上万的民众在这两座地堡（一座位于威廉斯堡，而另一座则位于圣保利区）中躲避盟军空袭。因配备了高射炮塔，该地堡也构成了德国战争机器的一部分。1947年，该建筑的内部在英国军队的一次控制爆破行动中遭到了毁灭性的破坏。8层楼中有6层出现倒塌，而其余的楼层也因为太过危险而不能进入。只有建筑的外壳——3m高的墙壁和4m厚的天花板几乎完好无损地保存了下来。在过去的60多年里，该建筑的进一步使用仅局限于几个毗邻的地区。如今，这座一直处于倒塌风险的地堡经过翻修，作为历史遗迹被保留下来，成为汉堡国际建筑展的一部分。

Courtesy of the architect

历史遗迹和能源地堡

"能源地堡"的屋顶和南边安装有太阳能套管，从远处清晰可见，成为易北岛新型再生能源供应道路上的重要里程碑。"能源地堡"以一种智能的方式综合利用太阳能、沼气、木屑以及邻近工业厂房产生的废热，为Reiherstieg区的大多数地方提供热能，同时也向汉堡配电网提供可再生能源电力。作为地区发电站，"能源地堡"代表了一种能够产生当地就业和收益的分散型能源政策。

智能网络中的创新技术

该项目的核心是在原防空地堡内建造的大型储热器。由于整合了不同类型的环保热能和电力装置，在今后的几年里，该建筑将转变成一座能够为方圆1.2平方千米的区域提供热能的"能源地堡"。该项目最具创新性的特征在于它的大型缓冲存储设施，预计其总容量可达2百万公升。该存储设施的热能输送来源于燃烧生物甲烷的热电联合装置、一个木材燃烧系统、一个太阳能保温装置，以及一家工业厂房产生的废热。

咖啡厅/展厅：地堡的历史

该悬臂式平台在30m的高度上围绕整栋建筑延伸开来，提供几乎可以鸟瞰整个汉堡的360°景观。建筑师在与Geschichtswerkstatt Wilhelmsburg（威廉斯堡&港口历史工作室）的紧密合作下，成功建立了一个侧重于展现防空地堡历史和其转变为"能源地堡"过程的令人印象深刻的展厅。由"Rotenhäuser场地"开始，参观者可以沿途学习二战期间地堡的建设情况以及由汉堡国际建筑展所开展的翻修设计，开启一场令人兴奋的发现之旅。

Energy Bunker

Wilhelmsburg's former air raid bunker has been transformed into a symbol of the "Renewable Wilhelmsburg" Climate Protection Concept. Having languished almost unused since the end of World War II, the monument has now been renovated during the IBA Hamburg and converted into a power plant using renewable forms of energy, with a large heat reservoir. This supplies the Reiherstieg District with climate-friendly heat, while feeding renewable power into the Hamburg's distribution grid.

The air raid bunker on Neuhöfer Strasse was built in 1943 to demonstrate the supposed valour of the home front. Thousands of people sought shelter from the allied bombing raids in such two bunkers, one in Wilhelmsburg and the other in St Pauli. With its flak towers, the bunker also formed part of the German war machine. In 1947 the interior of the building was completely destroyed by the British Army in a controlled demolition. Six of the

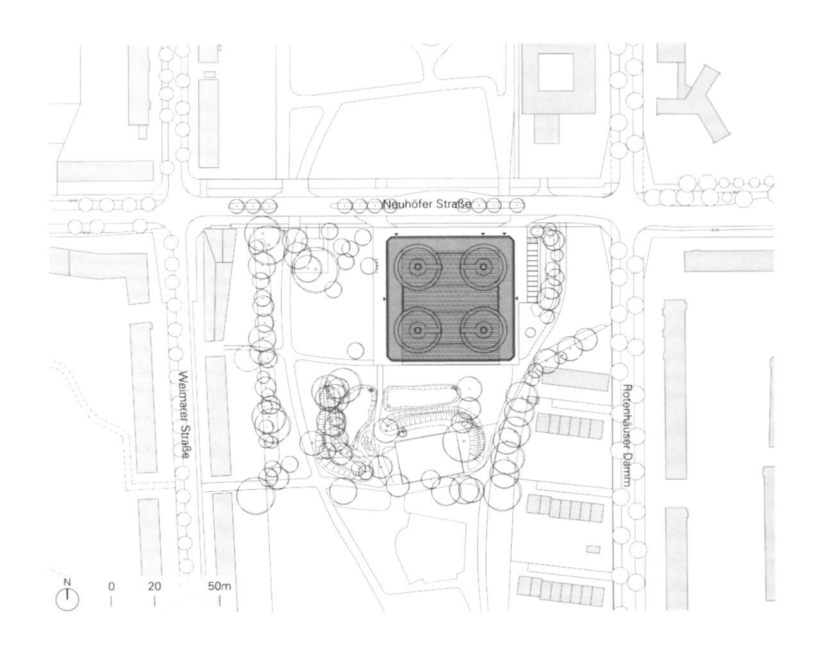

eight floors collapsed, and the rest was too dangerous to access. Only the outer shell of the structure, its walls up to three meters and its ceilings up to four metres thick, remained almost intact. For over sixty years, further use of the building was restricted to a few adjacent areas.

Now the bunker, which had been in danger of collapsing, is being renovated and preserved as a monument as part of the Internationale Bauausstellung IBA Hamburg.

Monument and Energy Bunker

With a solar casing on its roof and southern side, the "Energy Bunker", visible from a distance, marks an important milestone on the road towards supplying the Elbe Islands with renewable forms of energy. By using, in an intelligent way, a combination of solar energy, biogas, wood chips, and waste heat from a nearby industrial plant, the "Energy Bunker" is set to supply most of the Reiherstieg District with heat, while also feeding renewable power into the electricity grid. As a local power plant, the "Energy Bunker" represents a decentralised energy policy that creates local jobs and income.

Innovative Technology in a Smart Network

At the heart of the project is the large heat reservoir built inside

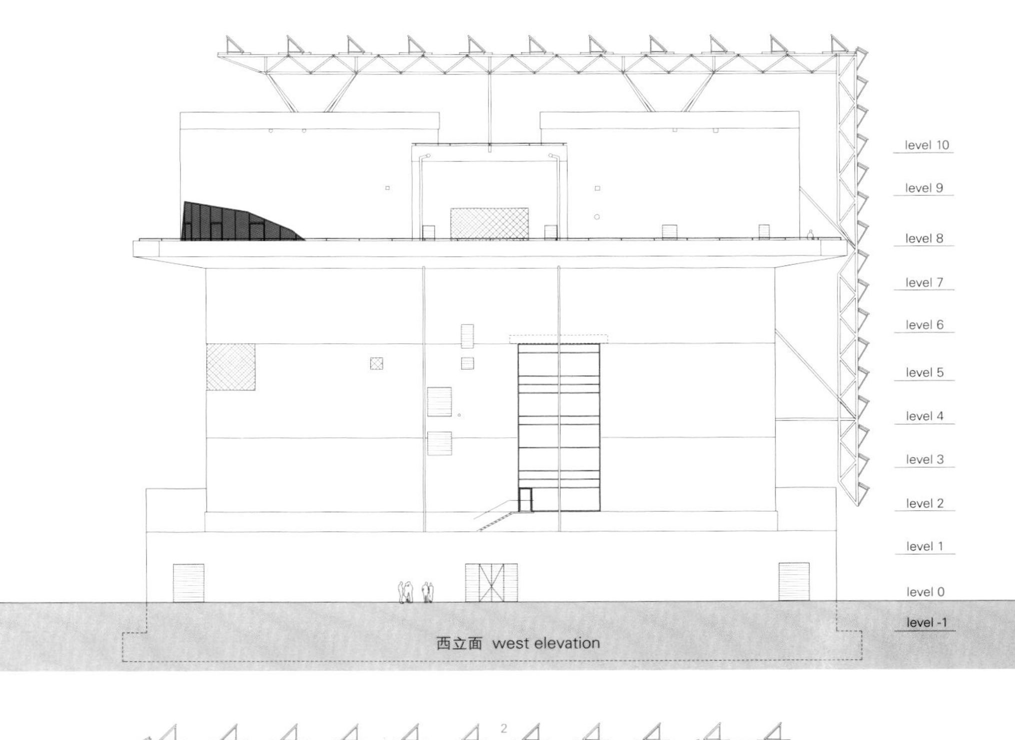

西立面 west elevation

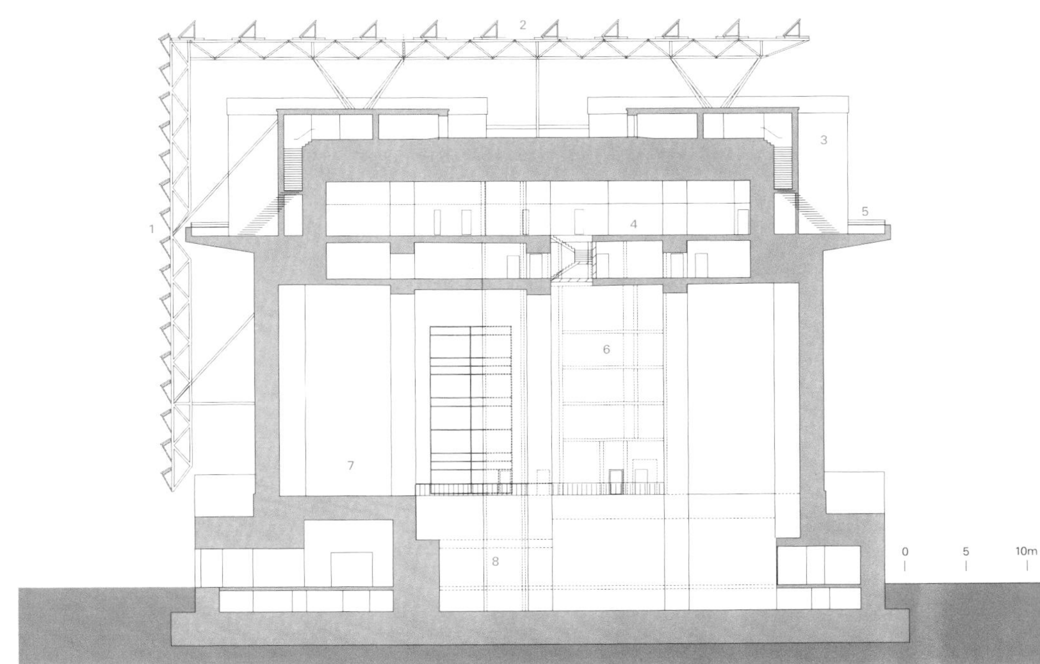

1 光伏系统 2 太阳能保温装置 3 防空炮塔 4 设有咖啡厅的8楼
5 悬臂板 6 缓冲存储设施 7 走廊 8 能源中心
1. photovoltaic system 2. solar thermal unit 3. anti-aircraft turret 4. level 8 with cafe
5. cantilever slab 6. buffer storage facility 7. gallery 8. energy center

A–A' 剖面图 section A-A'

the former air raid bunker. In the years to come it will transform the building into an "Energy Bunker" capable of supplying a district covering an area of more than 1.2 square kilometres, thanks to the integration of different types of environmentally friendly heat and electric power units. The project's most innovative feature is its large-scale buffer storage facility, with an expected total capacity of 2 million litres. This is fed by the heat from a biomethane-fired combined heat and power unit, a wood combustion system, and a solar thermal unit, as well as the waste heat from an industrial plant.

Cafe / Exhibition: History of the Bunker

This cantilevered platform, which runs around the whole building at a height of 30m, offers 360° views over almost all of Hamburg. An impressive exhibition focusing on the history of the air raid bunker and its transformation into the "Energy Bunker" has been created in close cooperation with the Geschichtswerkstatt Wilhelmsburg (Wilhelmsburg & Harbour History Workshop). Visitors embark on an exciting journey of discovery, beginning in "Rotenhäuser Feld", learning along the way about the construction of the bunker during World War II and its renovation by the IBA Hamburg.

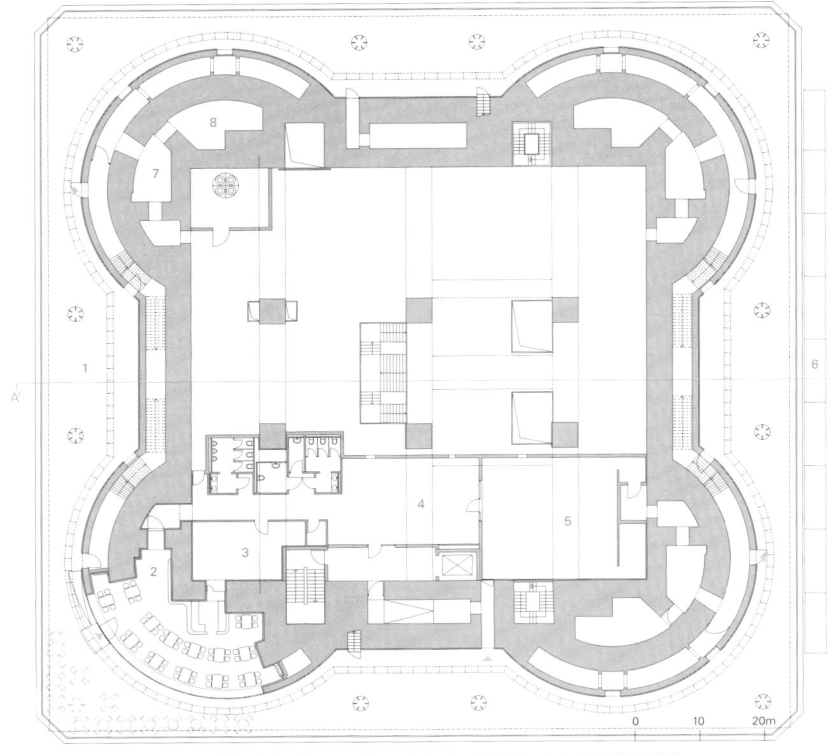

1 悬臂板 2 咖啡吧 3 厨房 4 地面 5 烘培设备 6 太阳能套管/钢结构 7 原气塞 8 原弹药存储库

1. cantilever slab 2. cafe bar 3. kitchen 4. floor 5. roasting facility 6. solar casing / steel structure 7. former gas lock 8. former munition storage

八层 eighth floor

项目名称：Energy Bunker
地点：Neuhöfer Straße 7, 21107 Hamburg, Germany
建筑师：Hegger Hegger Schleiff HHS Planer + Architekten AG
开发商：
building_IBA Hamburg GmbH,
energy supply_HAMBURG ENERGIE GmbH
景观建筑师：EGL
楼层面积：3,279m²
带有咖啡室的档案中心面积：1,420m²
能量中心面积：5,625m²
立面的太阳能板面积：Approx. 1,600m²(south side)
光伏系统面积：Approx. 1100m²(roof)
热水箱容量：2,000m³
总产量：heating for 3,000 and electricity for 1,000 households, CO_2 saving ~95%
高度：42m
甲方：IBA Hamburg GmbH
翻修前的竣工时间：1943
室内拆除时间：1947
改造时间：2010
竣工时间：2013
摄影师：
©Bernadette Grimmenstein (courtesy of the architect) - p.46~47, p.48[bottom], p.49, p.52
©Martin Kunze (courtesy of the architect) - p.48[middle], p.50~51 p.53

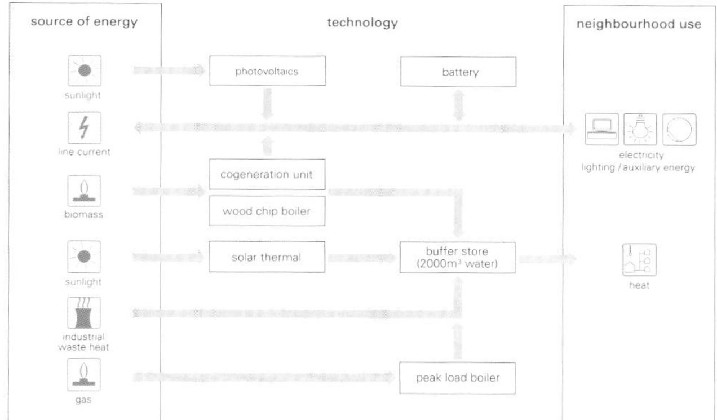

能源理念 energy concept

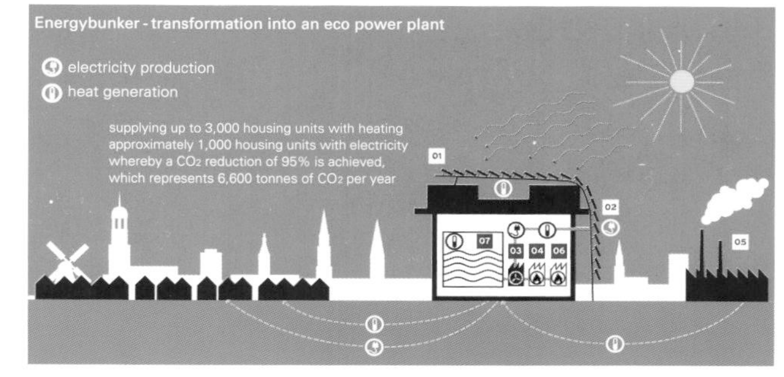

1. A rooftop solar thermal unit generates heat from the sun.
2. A photovoltaic system on the south-facing facade produces electricity.
3. A biogas chp plant produces electricity and heat.
4. A woodchip heating system generates heat.
5. Waste heat from a neighbouring industrial plant is fed into a storage bunker and fed into the heating grid.
6. A peak load power plant ensures a steady supply of heat by covering load peaks.
7. A heat storage unit bunkers surplus heat, reacts to periods of peak demand and maintains supply.

拆除建筑的剖面图 demolition section

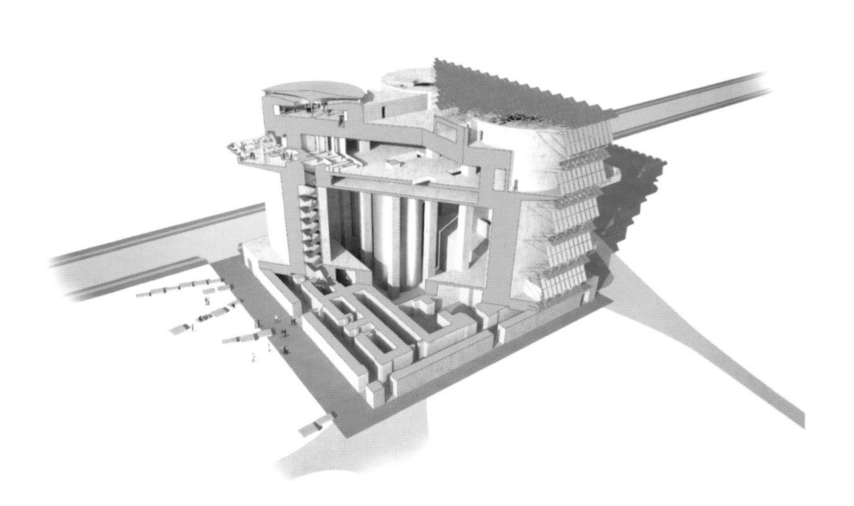

利用太阳能电池板的零能耗学校

Mikou Design Studio

码头学校位于圣图安市重要的城市区域——Zac des码头混合开发区的中央位置，该区域是当地可持续城市发展的典范之作。

虽然学校周围布满了高层的办公和住宅大楼，但因临近建筑的窗口都面对其屋顶这样的特殊设计，仍然使其成为城市建筑中无法忽视的一部分，而这样的设计也成为该项目的一个基本特征。

建筑师设计这栋零能源消耗设施的目的在于使之成为Zac des码头可持续发展项目的标志性建筑，同时由于该校的地理位置，以及为孩子们提供了舒适的学校内部设计，尤其是学校的操场和花园的设计，也使其成为该地区的强大的地标性建筑。而配有太阳能光伏板的区域将建筑美学充分地融入其中，并且人们可以从主街道看到这处区域，使学校具有浓郁的教育机构特性。

该建筑的地理位置有助于所有的教室全部朝南设计，操场亦是如此，这样一来就可以尽可能地利用被动式太阳能。同时，如此的空间排列也尽可能地扩大了南面配有太阳能光伏板的区域面积，使其融入带顶篷操场区域的建筑中。

因此，该项目的设计方案是在主街东面建造一个阶梯状的建筑体量，通过大型顶篷延伸开来，并在场地对角线的交叉条状结构处进行折叠，以使建筑体量面向南方。

室内花园将这些呈阶梯式的交叉条状结构分割开来，并提供了清晰的东西方向的校园景观，同时让阳光照射到各个教学区域，且具有清晰的辨识性，也为带顶篷的内部露台带来了平和与安静。

校园面朝南方，在明亮的露台上设计有一系列的花园和建筑体量，它们逐渐地向下延伸，释放人们的视野，并使阳光最大限度地照进这座建筑。

Zero Energy School in Solar Panels

The Docks School in Saint Ouen is in a strategic urban location, in the middle of the Zac des Docks mixed development area, which is an exemplary case of sustainable urban development.

Located in the middle of an urban complex composed mainly of high-rise office blocks and housing, it will also be visible because of its roof, a fundamental feature of the project onto which the openings of neighbouring buildings face.

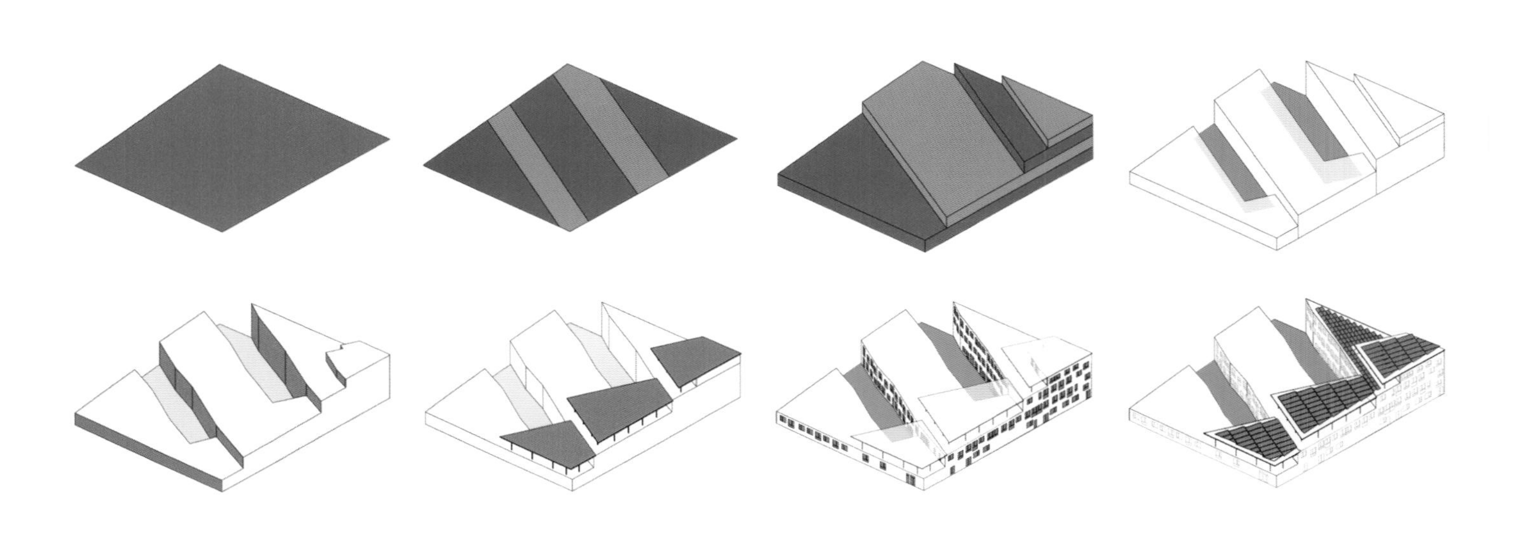

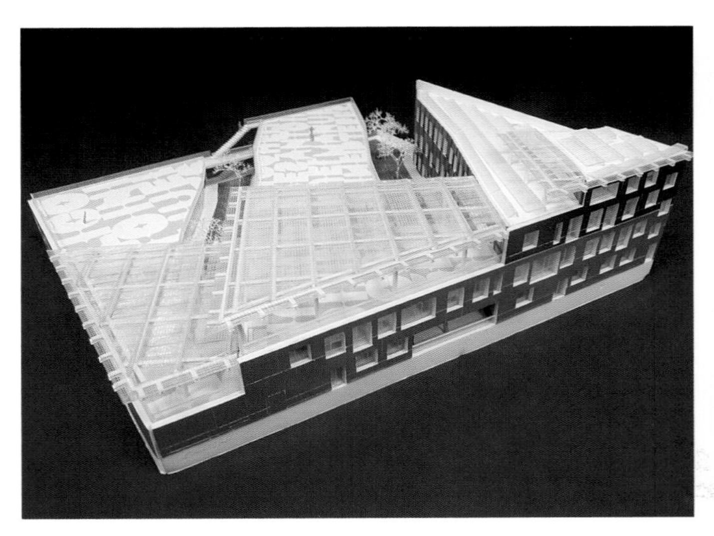

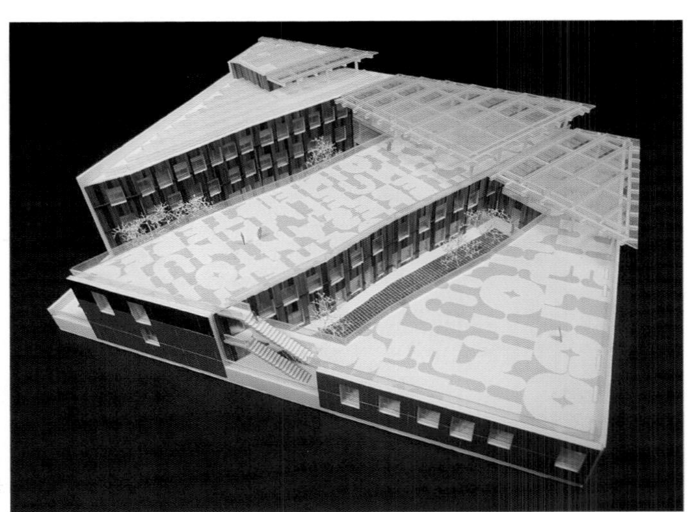

1 露台　2 幼儿活动场地　3 覆盖幼儿活动庭院的藤架结构　4 洗手间
5 教室　6 幼儿活动庭院　7 休息室　8 运动室　9 讨论室　10 主教室
1. patio 2. infant court 3. pergola structure covering infant yard 4. WC 5. classroom
6. infant yard 7. restroom 8. motricity room 9. group study room 10. master room
二层 second floor

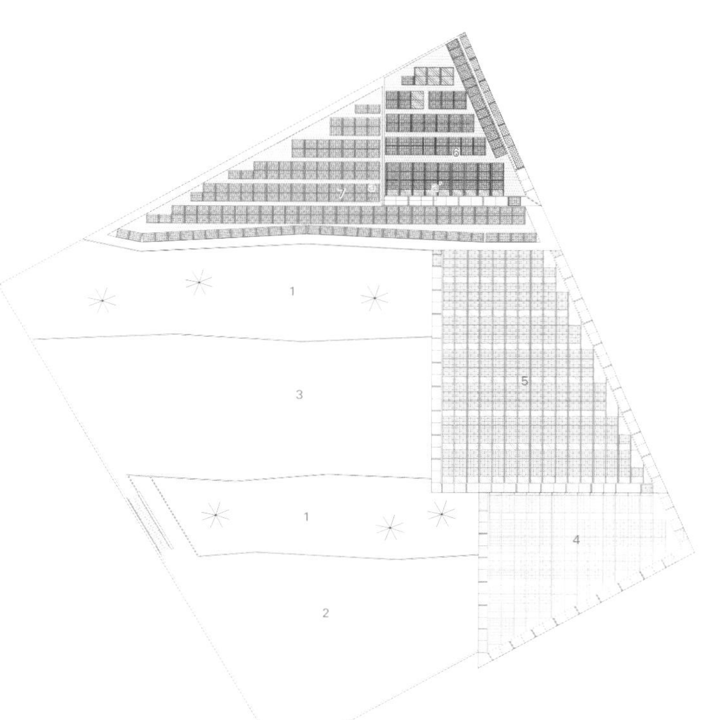

1 露台　2 幼儿活动场地　3 小学生活动场地　4 覆盖幼儿活动庭院的藤架结构
5 覆盖小学生活动场地的藤架结构　6 平台屋顶　7 太阳能光伏板
1. patio 2. infant court 3. elementary court 4. pergola structure covering infant yard
5. pergola structure covering elementary yard 6. terrace roof 7. photovoltaic panels
屋顶层 roof

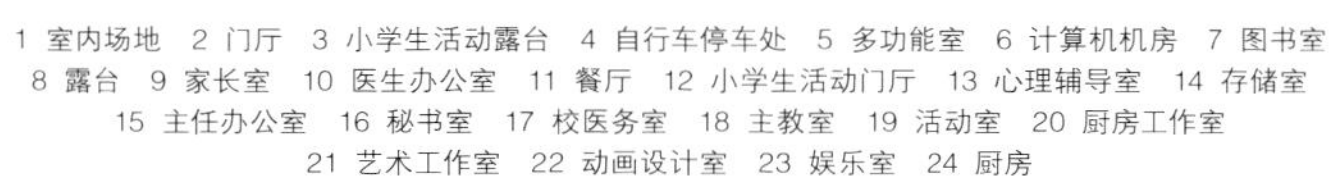

1 室内场地　2 门厅　3 小学生活动露台　4 自行车停车处　5 多功能室　6 计算机机房　7 图书室
8 露台　9 家长室　10 医生办公室　11 餐厅　12 小学生活动门厅　13 心理辅导室　14 存储室
15 主任办公室　16 秘书室　17 校医务室　18 主教室　19 活动室　20 厨房工作室
21 艺术工作室　22 动画设计室　23 娱乐室　24 厨房
1. covered court 2. hall 3. elementary patio 4. bike 5. multipurpose room 6. computer room
7. library 8. patio 9. parents room 10. doctor 11.dining room 12. elementary hall 13. psychologist
14. deposit 15. director 16. secretary 17. infirmary 18. master room 19. activity room
20. kitchen atelier 21. art atelier 22. animator room 23. game room 24. kitchen
一层 first floor

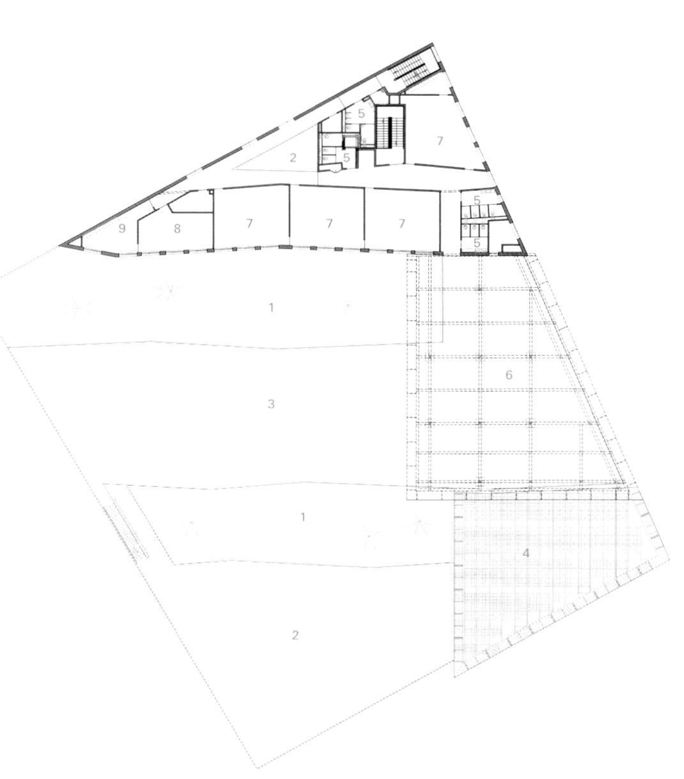

1 露台　2 幼儿活动场地　3 小学生活动场地　4 覆盖幼儿活动庭院的藤架结构　5 洗手间
6 覆盖小学生活动场地的藤架结构　7 教室　8 讨论室　9 主教室
1. patio 2. infant court 3. elementary court 4. pergola structure covering infant yard 5. WC
6. pergola structure covering elementary yard 7. class room 8. group study room 9. master room
三层 third floor

东北立面 north-east elevation

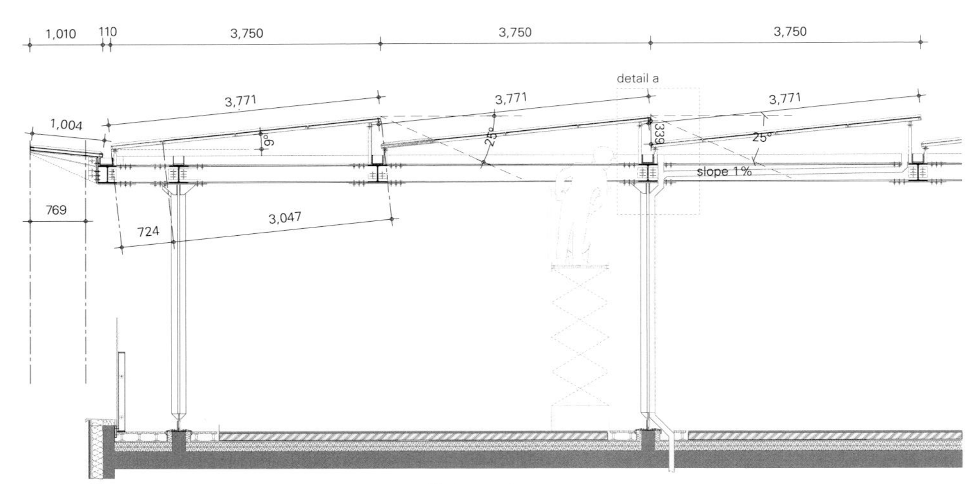

详图1 detail 1

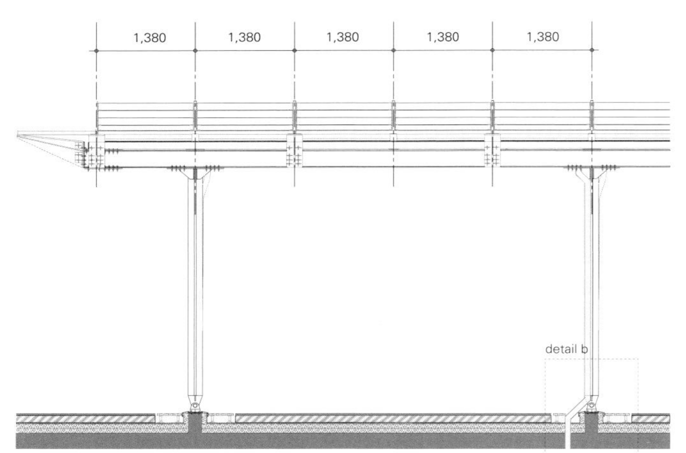

a–a' 剖面图 section a-a'

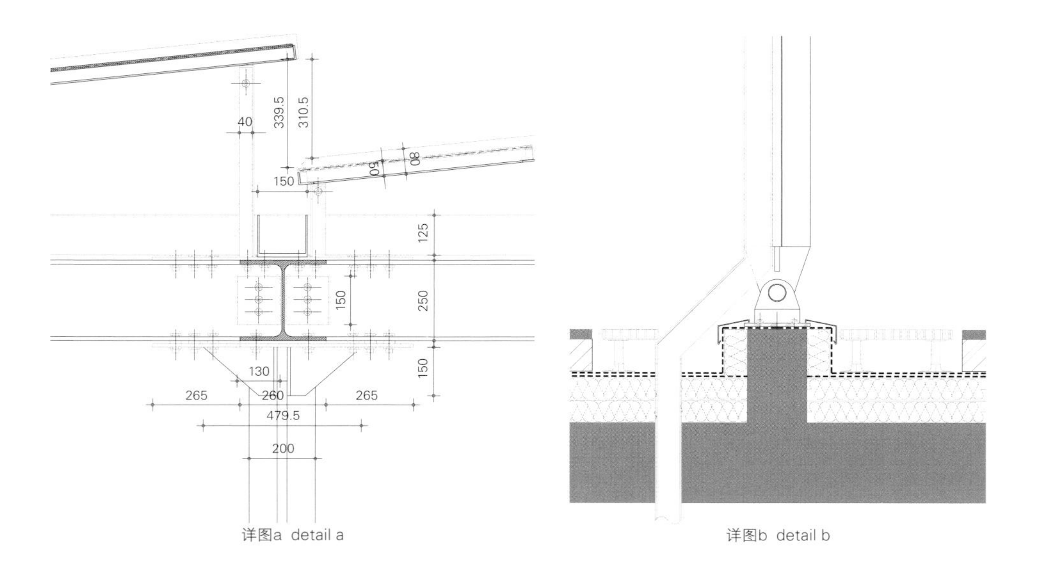

详图a detail a

详图b detail b

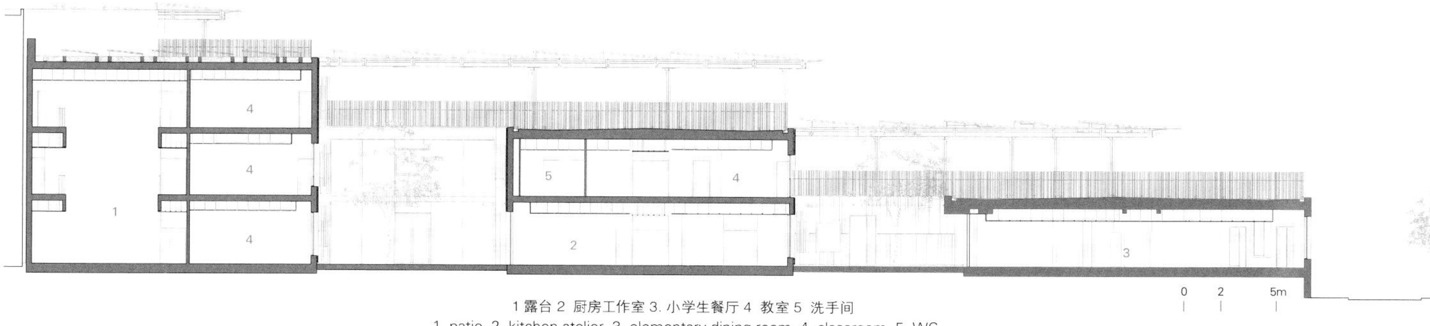

1 露台 2 厨房工作室 3. 小学生餐厅 4 教室 5 洗手间
1. patio 2. kitchen atelier 3. elementary dining room 4. classroom 5. WC
A–A' 剖面图 section A-A'

The architects designed this amenity to consume zero energy in order to be emblematic of the sustainable development of the Zac des Docks project and to be a strong architectural landmark in its neighborhood, which is exemplary by its choice of siting, by the interior comfort provided for the children – particularly in the design of the school playgrounds and gardens – and by the treatment of the areas of photovoltaic panels, which are integrated into the architecture and are visible from the main street, giving the school a strong educational identity.

The building's siting facilitated south-facing orientation of all the classrooms and playgrounds in order to make the greatest possible use of passive solar energy. This spatial disposition also made it possible to increase the surface areas on the south required for the photovoltaic panels which were integrated into the architecture of the covered playground areas.

Therefore the scheme is in the form of a mass built in stepped tiers on the east, on the main street, which is extended by large canopies and which folds in crosswise strips on the site's diagonal to face southwards.

These stepped crosswise strips are separated by internal gardens which open wide east-west transparent views into the school while allowing clear identification of the various teaching areas open to the light and the peace and quietness on sheltered internal patios.

Facing southwards, the school is arranged in a succession of gardens and volumes in brightly-lit terraces which gradually descend to free the view and to let in maximum sunshine.

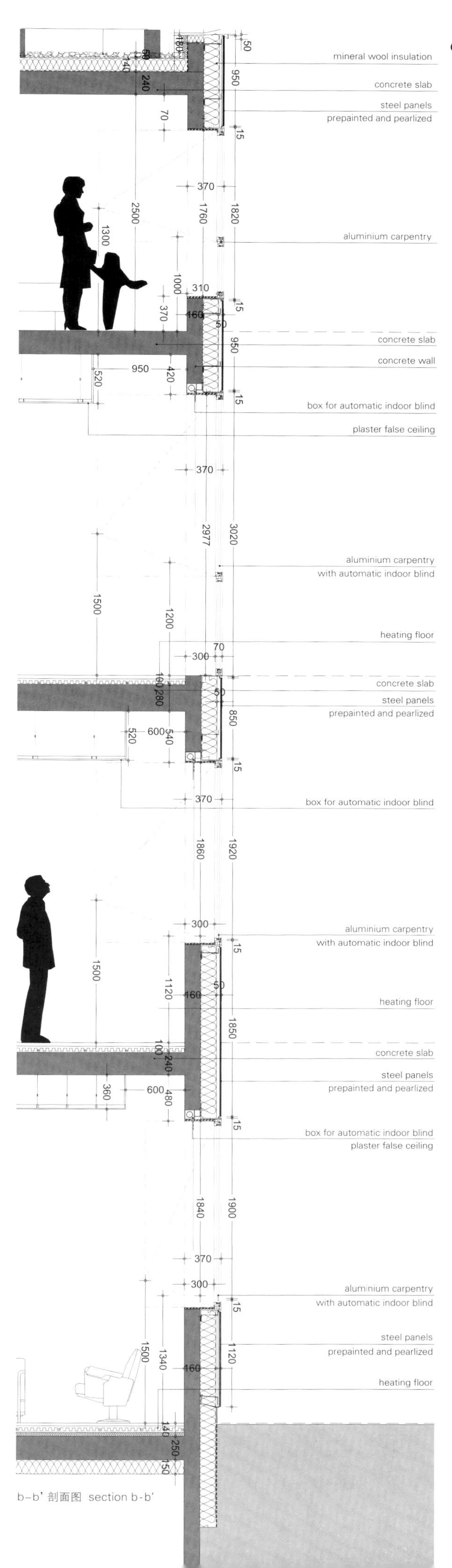

b–b' 剖面图 section b-b'

项目名称：Zero Energy School
地点：Saint Ouen, France
建筑师：Mikou Design Studio
结构和服务工程师：Berim
工料测量师：Sletec
景观建筑师：Mikou Design Studio
主要承包商：Urbaine Des Travaux / Groupe Fayat
甲方：Sequano Amenagement and City Of Saint Ouen
功能：Elementary school, preschool, sport and recreation center, restaurant, administration
用地面积：4,820m²
竣工时间：2013.10
摄影师：©Florian Kleinefenn(courtesy of the architect)

©Helmut Lunghammer

格拉茨的斯巴克埃森–霍费尔储蓄银行

Szyszkowitz-Kowalski + Partner ZT GmbH

西南立面 south-west elevation

这座扩建的储蓄银行坐落于格拉茨市区的中心地段，位于斯巴克埃森广场和安德烈亚斯·霍费尔广场之间，紧邻安德烈亚斯·霍费尔广场，其隐蔽性极佳的室内场地占据了街区的大部分面积。

从总体上看，这座建筑在周边历史中心（从不放弃在其外围延续生存力）代表了前瞻性的深入发展的承诺。

几年前，它的立面进行了翻新，从内部结构来看，现在被建成了一座功能齐全的主楼。所有楼层都围绕整栋大楼布局，形成开放、流通的环形结构。

这座崭新的综合设施位于且面向安德烈亚斯·霍费尔广场，由七个楼层组成。这栋建筑取代了街区里一栋破旧的非列管建筑，现在已被拆毁，以及街区内的一座4～6层的U形建筑。所有边界形成了一个室内庭院，庭院由位于四层的大型可延伸的水平向板条进行遮阴。庭院的设计理念类似于空中花园，采用倒置的阶梯状金字塔结构，在上面种植上郁郁葱葱的植物。立面将八棵繁茂的大树融在一起，且屋顶露台还有四棵，几片草地和一层的游泳池都为整栋建筑营造了一种特殊的氛围。

新建的中央庭院的立面表面主要是由玻璃和钢结构构成的，或倾斜或呈阶梯状，确保了每一楼层可以最大限度地得到光照，也可以进入每层或狭窄或宽阔的露台区域。

建筑师尤为注意面向安德烈亚斯·霍费尔广场的唯一真正的外立面：它反映了面向室外上空开放的室内庭院的线条，尽管其自身的外形连接着相邻建筑的不同高度，但是通过其层次和后退距离，还是创造了一个整体的形象，不过这一切都归功于建筑的弧形屋顶。突出的石匠工艺打造了全新的、仅有的外观。玻璃板条围合的圆锥形区域暗示了内部庭院的金字塔结构，同时避免夏季温度过高。

替代能源理念的特点是从两个深井取用地下水，并利用其水温。每个楼层的天花板里安装了带有额外快速反应表面的热激活系统，可以进行冬季供暖和夏季制冷。此外，庭院被大型移动遮阳板条遮盖，板条可以移动，然后停留在圆弧形屋顶，并在低光照时段充当反光篷。这种设计保证了整栋建筑里的自然光线和通风得到了最大限度的利用。另外，建筑格局多以狭长结构为主，例如小进深结构确保了良好的照明、通风和透明度，且贯穿了整座建筑。

该建筑的楼层平面鼓励设计一个开放且流动的办公室布局和独立的办公环境。除了氛围以外，特别值得关注的是高效的降低噪声系统和特殊的照明系统，它们可以由个人控制。位于地面层、新室内庭院一侧的休息接待室面向现存的主建筑开放，并与之相连。

项目名称：Sparkassenhöfe Graz
项目地点：Andreas-Hofer-Platz 9, A - 8010 Graz, Austria
建筑师：Szyszkowitz-Kowalski + Partner ZT GmbH
项目团队：Transsolar Energietechnik GmbH, Tragwerksplanung ZT
甲方：Steiermärkische Bank, Sparkassen AG
用地面积：927m² / 可用楼层面积：10,230m²
设计时间：2008 / 施工时间：2009 / 竣工时间：2011
摄影师：
©Angelo Kaunat (courtesy of the architect) - p.69, p.71, p.72, p.73, p.74, p.75
©Stadtvermessung Graz (courtesy of the architect) - p.67

Sparkassenhöfe Graz

The extended building of the Saving Bank is situated in the middle of central Graz between Sparkassenplatz and Andreas-Hofer-Platz, with the building abutting directly on Andreas-Hofer-Platz and the well-hidden interior area occupying the majority of the block.

As a whole, it represents a commitment to forward-looking further development within historic centers that have not relinquished their viability to the periphery.

Internally, it constitutes a functional whole with the "main building" that was given a new facade a few years ago. All storeys are arranged around the entire house as a kind of open, flowing ring connection.

The whole new complex consists of the seven-storey section bordering on and facing Andreas-Hofer-Platz. It replaced a decrepit unlisted building, which is now demolished, and a four-to-six-storey U-shaped building inside the block. These measures create a central interior courtyard shaded by large extendable horizontal slats on the fourth floor. The courtyard takes the form of an in-

四层 fourth floor

三层 third floor

一层 first floor

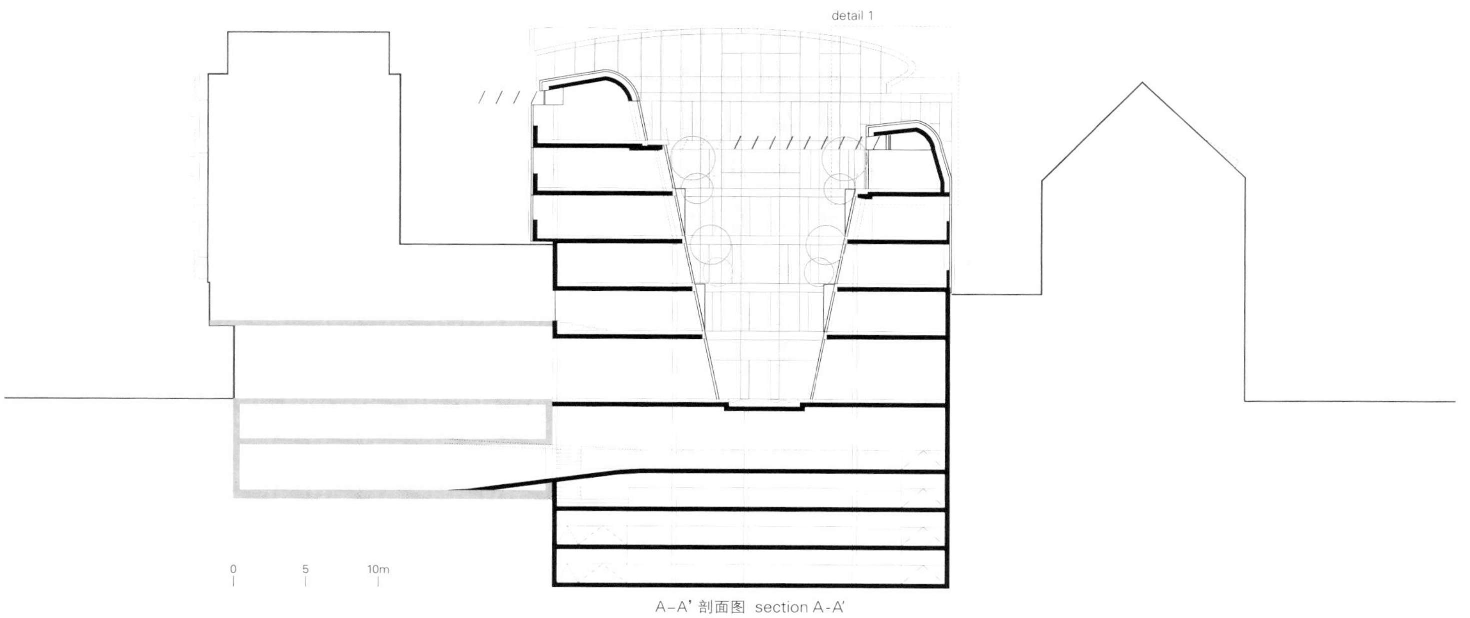

A–A' 剖面图 section A-A'

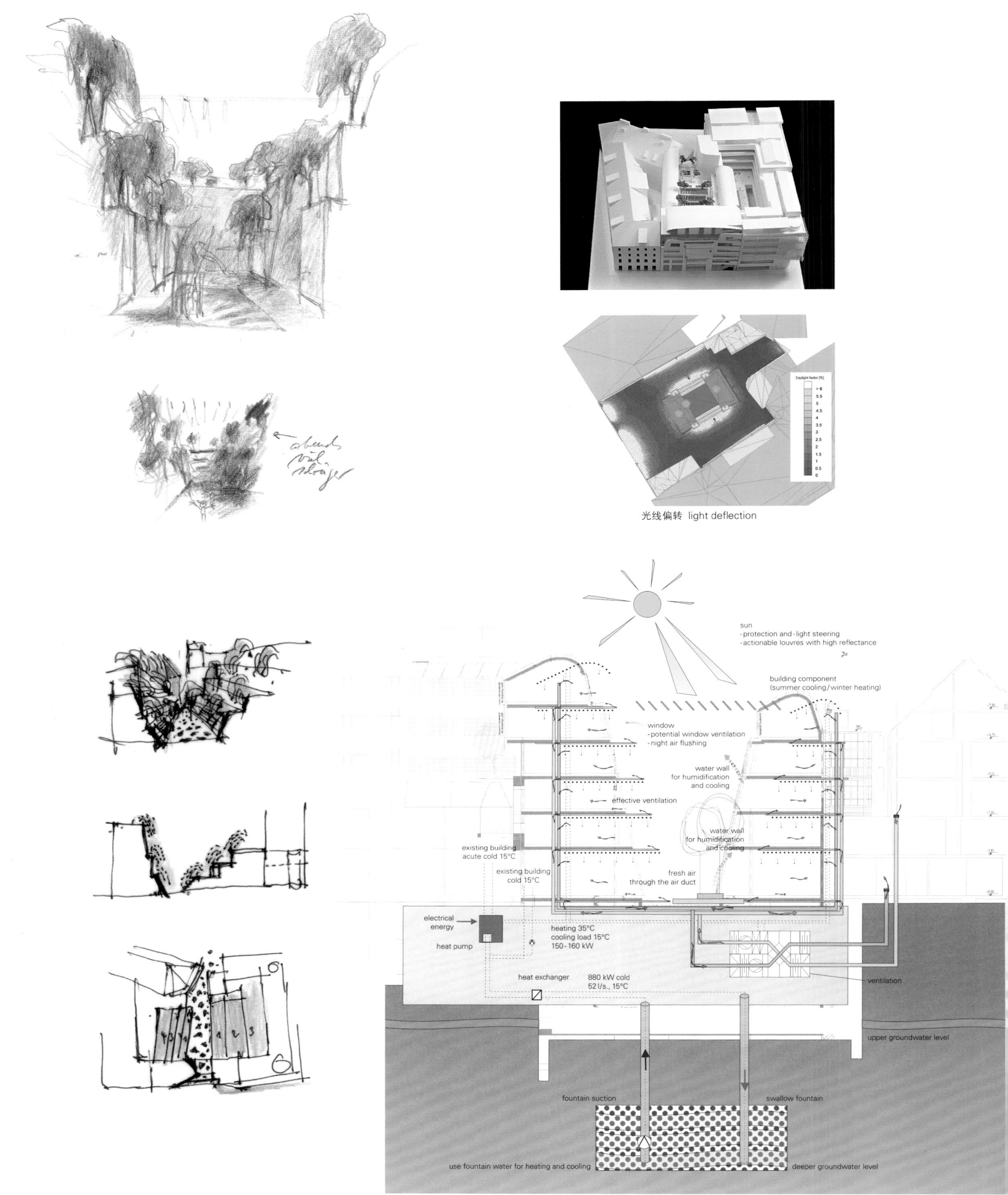

光线偏转 light deflection

能源理念 energy concept

verted stepped pyramid with lush plantings suggesting the idea of hanging gardens. Eight voluminous tree troughs are integrated into the facade, with another four on the roof terrace; a number of grassy fields, and a pool on the ground floor lend the building a special atmosphere.
The facade surfaces of the new central courtyard consist chiefly of glass and steel structures, inclined or stepped, and thus ensure maximum lighting but also access to the narrow and in some cases wide terraces formulated on every storey.
Special attention is given to the only real exterior facade, the one facing Andreas-Hofer-Platz: It reflects the lines of the interior courtyard that opens out upwards towards the outside, and although its own form connects the different heights of the neighboring buildings to the left and right, it also creats an overall impression by means of its layers and set-backs, but above all thanks to the curved roof. It is the only new facade with a notable amount of masonry. A conic field of glass slats alludes to the pyramid form of the interior courtyard and provides protection from overheating in summer.
An alternative energy concept features two deep wells to take advantage of groundwater temperature, with the aid of a thermoactive system installed in each of the storey's ceilings that provides heating in winter and cooling in summer, with additional quick-reaction surfaces on the ceilings. In addition, the courtyard is shaded by large, moving sunshade slats, which can be parked in the compass roofs and used as light canopies during hours of low light. This arrangement ensures that natural light and ventilation in the entire complex are used to their best advantage. Additionally, the architecture consists of narrow structures, i.e. short room depths that already ensure good lighting, ventilation and transparency throughout the building complex.
The floor plan encourages an open, flowing organization of offices and individual working situations. In addition to the atmosphere, special attention is given to highly effective noise reduction and special lighting that can be individually controlled.
The reception and lounge area alongside the new interior courtyard is situated on the ground floor area and openly connected to the existing "main building".

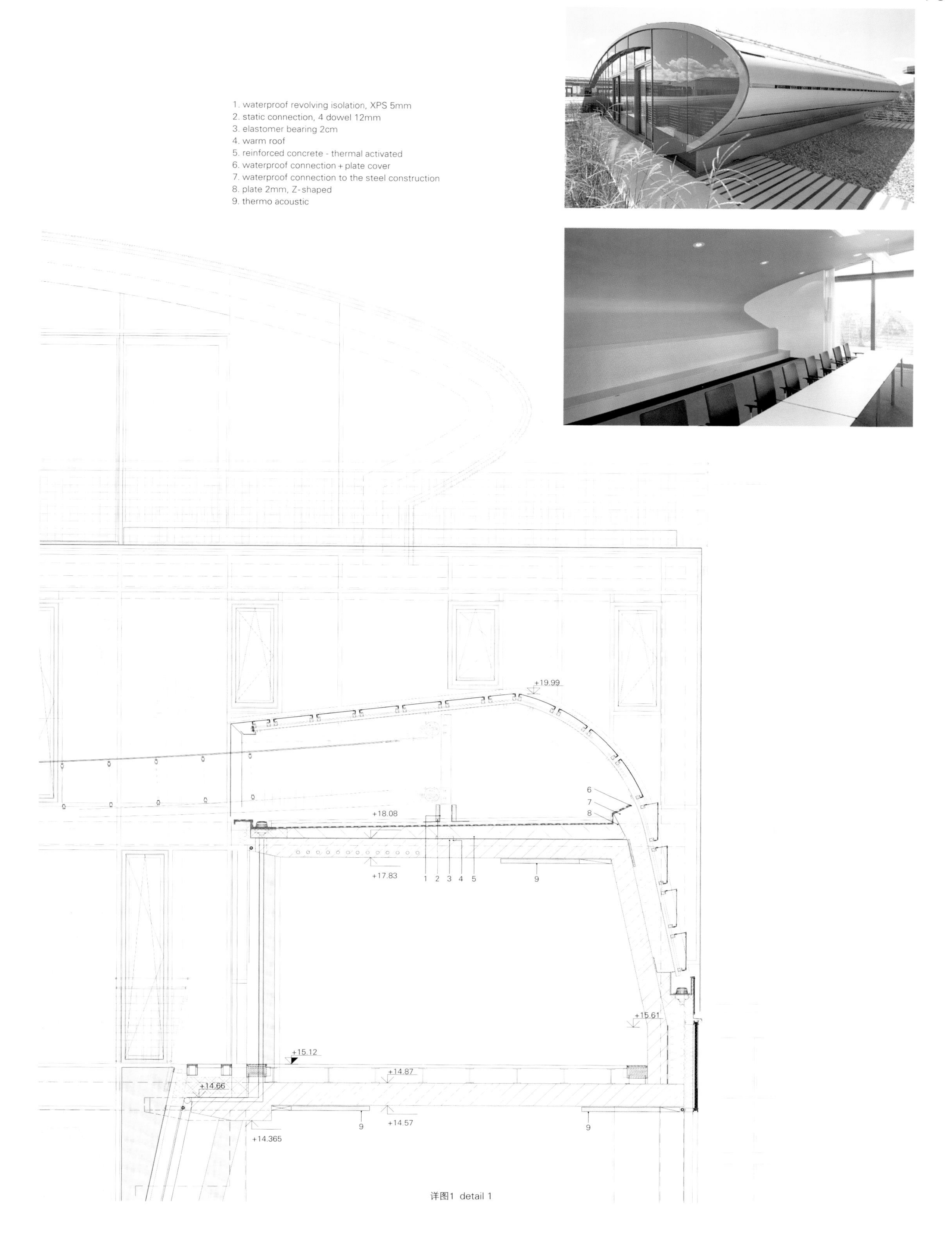
1. waterproof revolving isolation, XPS 5mm
2. static connection, 4 dowel 12mm
3. elastomer bearing 2cm
4. warm roof
5. reinforced concrete - thermal activated
6. waterproof connection + plate cover
7. waterproof connection to the steel construction
8. plate 2mm, Z-shaped
9. thermo acoustic
+19.99
+18.08
+17.83
+15.61
+15.12
+14.87
+14.66
+14.57
+14.365

详图1 detail 1

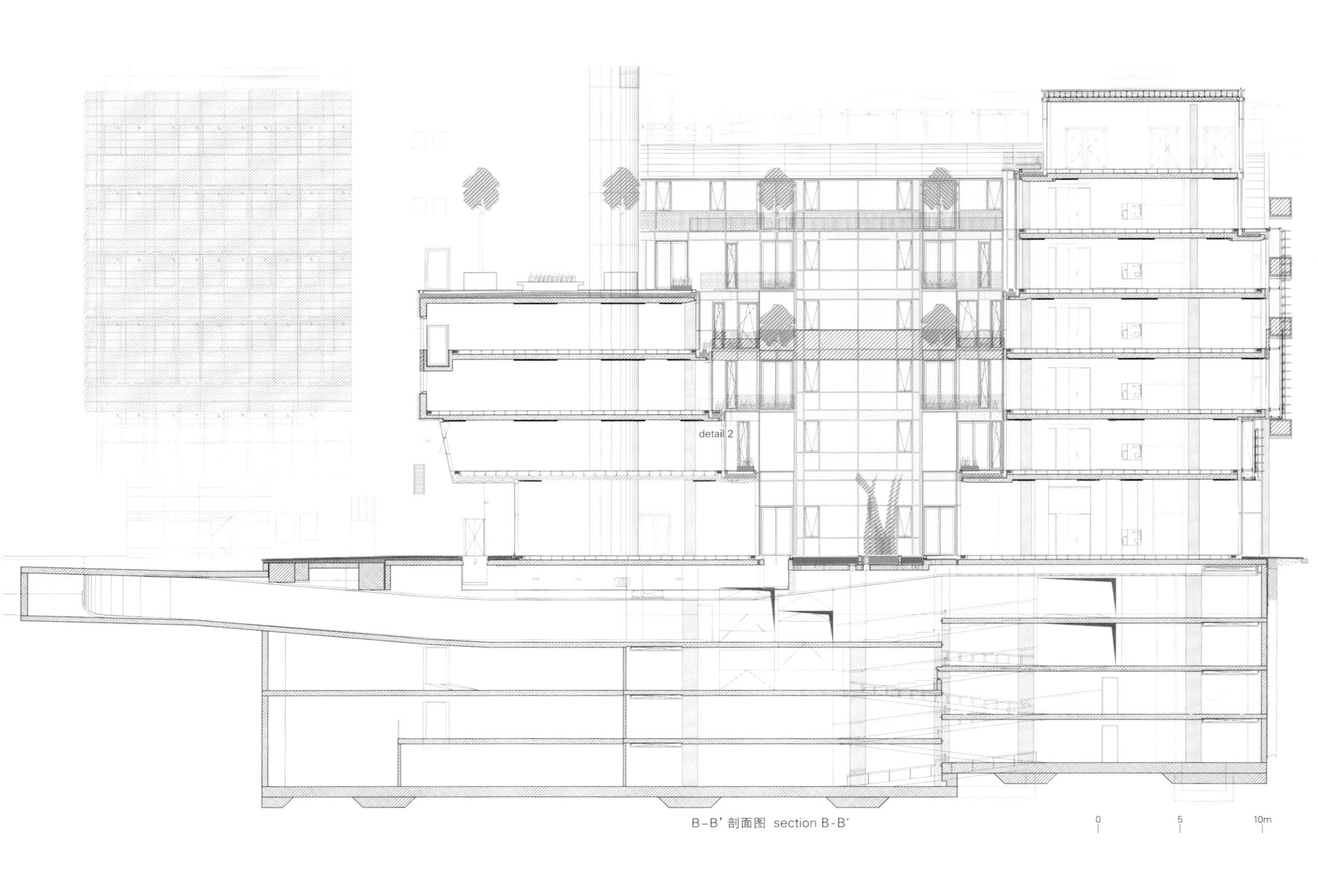

B-B' 剖面图 section B-B'

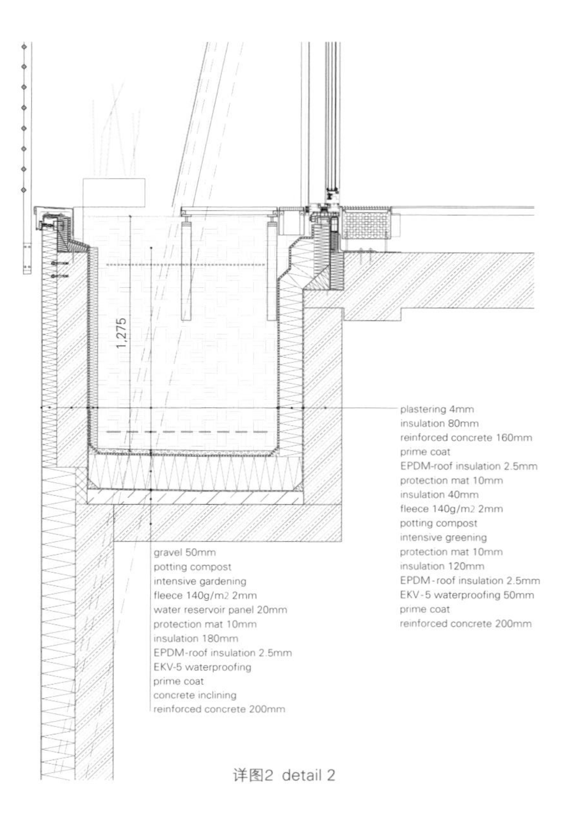

详图2 detail 2

特兰托科学博物馆

Renzo Piano Building Workshop

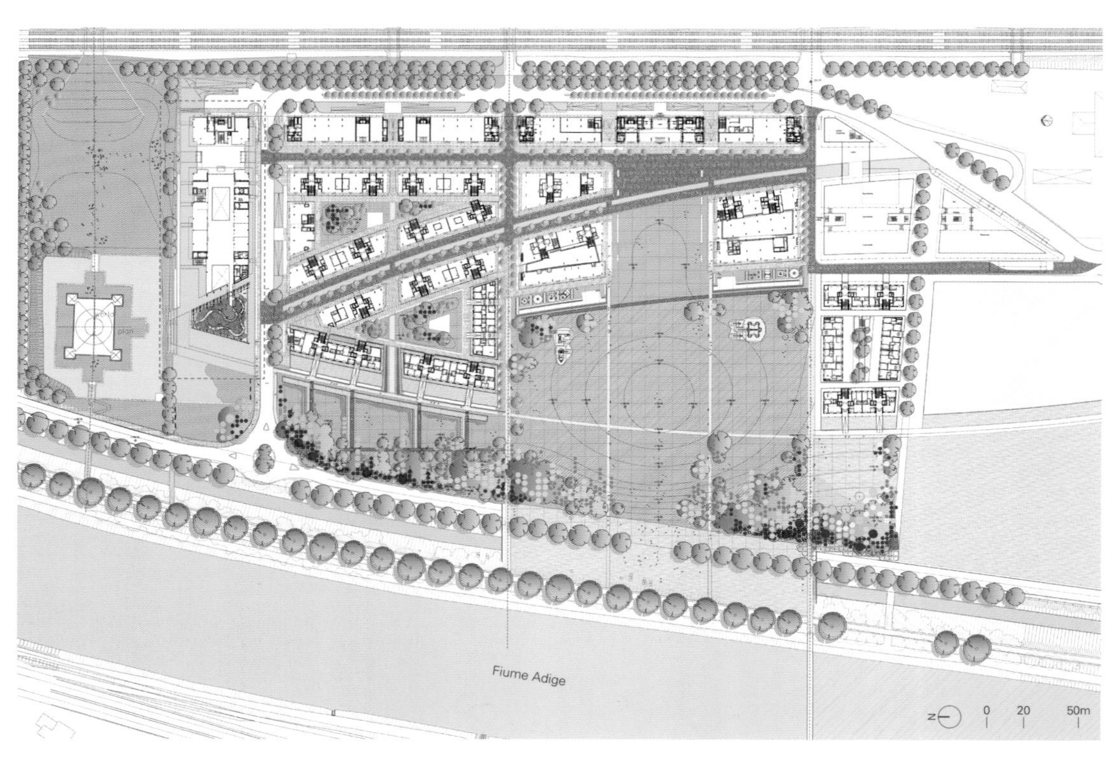
Fiume Adige
0 20 50m

新建的特兰托科学博物馆位于预计成为Ex-Michelin产业区新区的北部，坐落在人们所熟知的A街区，即连接该地区的高端活动和最大公众利益职能的人行道一端。同时它也紧挨着新建的公共公园和阿贝尔宫殿酒店，拥有多元化的关系。

这座建筑的理念是建立在对灵活性的要求，以及对文化项目本身的科学内涵的一致且严格的认同的基础上的一个完美折衷方案。博物馆宏伟的展览主题可以从形式和结构体量本身来识别，所有一切都保留了现代博物馆的典型灵活布局。

除了对博物馆的科学内涵进行体量解读，其建筑设计也受到博物馆与周围环境的关系的影响：更准确地说，是与新区（包括公园、河流和阿贝尔宫殿酒店）的关系的影响。正因如此，所有的投入实际上已经初具规模，这一切都要归功于对特殊建筑元素更加清晰的定义，最重要的是，这些元素弥补了街区其他地方的不足，尤其是第三产业，即居家和商业的作用。

整栋建筑是由一系列的空间和体量组成，实体和上空体量依附或漂浮在大型水体之上，使光与影的效果和动感更加多元化。宽阔的屋顶层把整个结构很好地组合在一起，与其形式和谐一致，即使从外面看这栋建筑也非常容易给人们留下深刻的印象。从东边来看，第一个结构不对大众开放，因为他们多为行政办公室和研究室、科学实验室和现场工作人员的辅助空间。

接下来，我们来到了大厅。它与整个街区的主轴平行排成一列，并且北贯整座建筑的进深，在此可以俯瞰阿贝尔宫殿酒店外面的公园区域。

山脉和冰川的科学主题通过一系列的展览空间得以展现，这些空间从地下室层面逐渐升起，上升到几乎穿破屋顶的层面，为大家开辟了一处从环境中崛起的观察点，使人们感受到一种仿真的体验。这种体验通过两层或三层的丰富的展览空间得以突出，且天花板也足够高，非常有利于设置大布景和背景。

建筑的外形，或者说"热带雨林"的功能也都用来定义室内空间和功能。实际上，这栋建筑展示了一间大型热带温室，在一年中的某些时期，它有能力与一些特定的展位甚至是室外空间（水景、照明和绿化在定义参观者周边的自然环境时扮演着至关重要的角色）建立功能性的关系。

沿着展区设置的一系列地面建筑为大众提供了教育和实验服务，从而推动了各个单独主题的交互式体验。

Trento Science Museum

The new Trento Science Museum is located in the northern portion of the new district foreseen for the Ex-Michelin area, and is housed what is known as the A-Block, situated at the end of the main pedestrian route that connects the area's higher-end activities with the functions of the greatest public interest. It is also located in close proximity to the new public park and Palazzo delle Albere, with which it will boast a productive relationship.

The idea was based on establishing a perfect compromise between the need for flexibility and the desire for a precise and consistent response to the scientific content of the cultural project itself. The museum's magnificent exhibition themes can even be recognized in the forms and volumes of the structure itself, all while maintaining the flexible layout typical of a more modern museum.

In addition to the volumetric interpretation of the museum's scientific contents, the architectural design has also been dictated by the museum's relationship with its surrounding environment: or rather the new district, including the park, the river and Palazzo delle Albere. Thus, all these inputs have physically taken shape

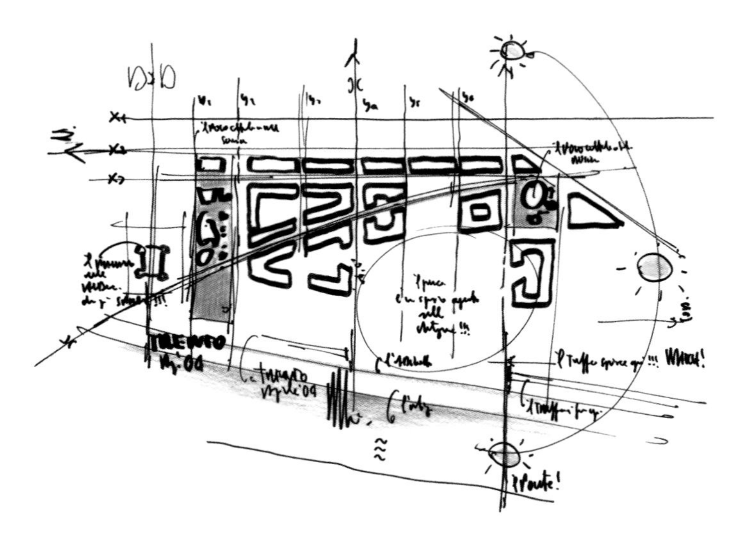

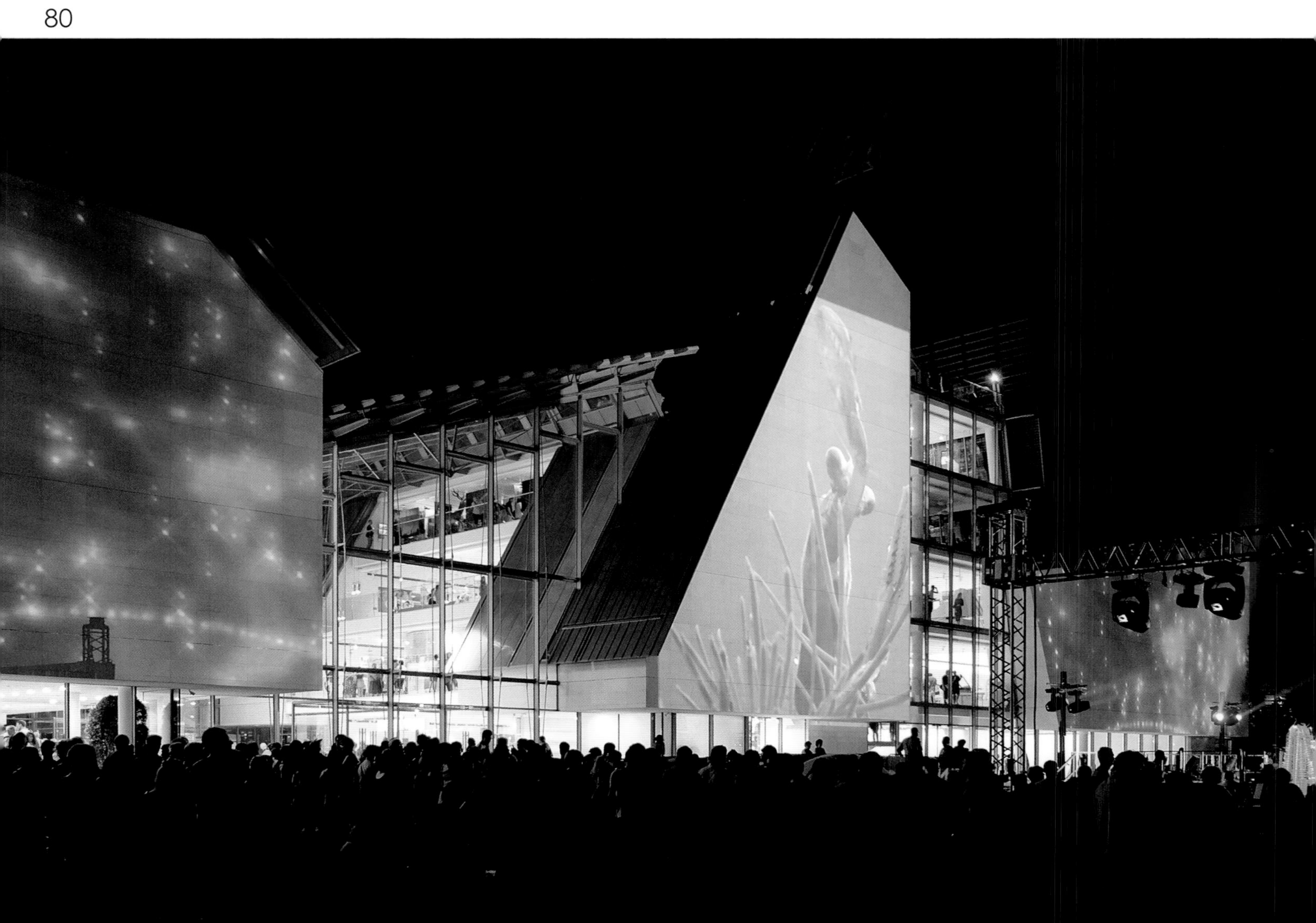

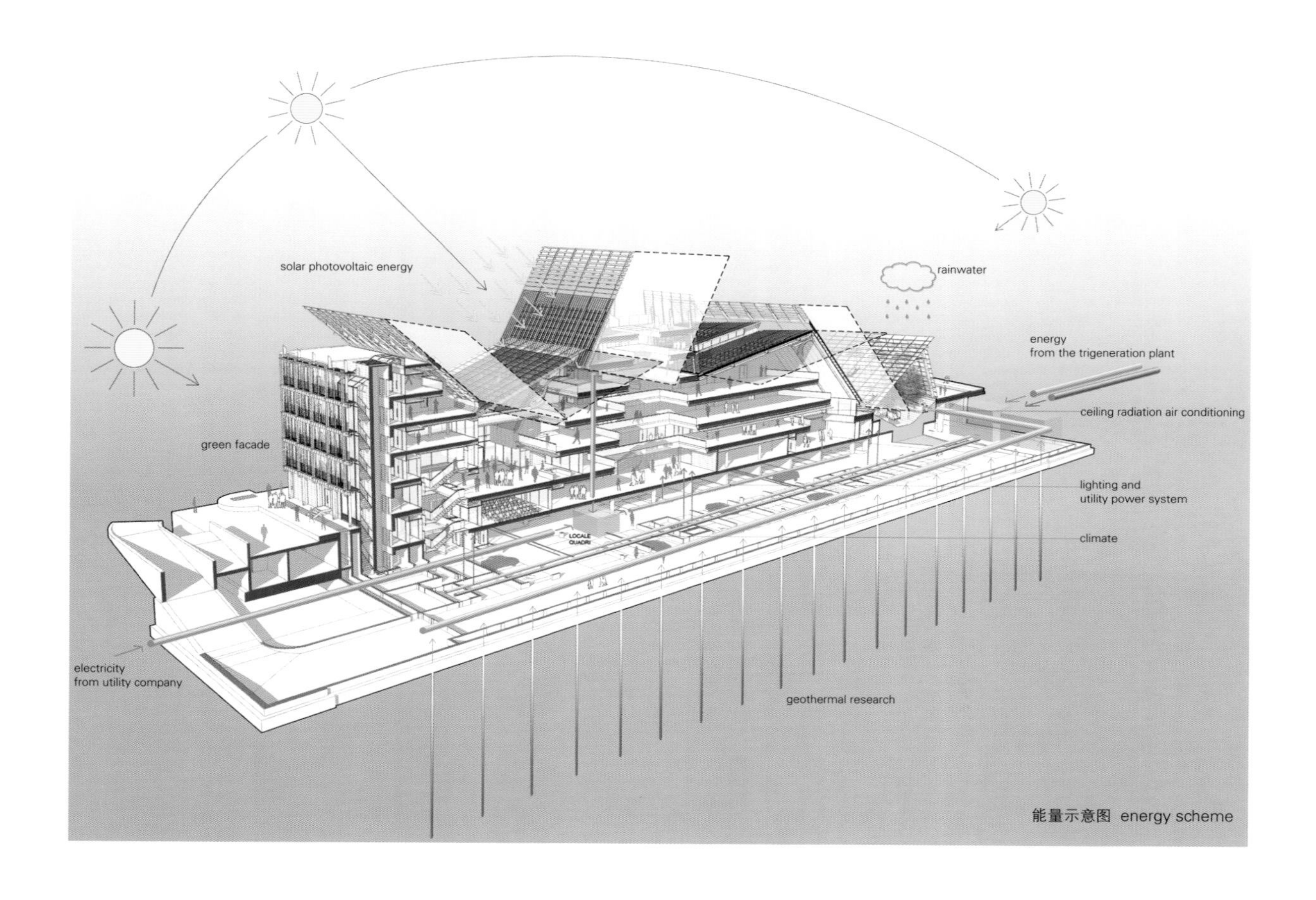

能量示意图 energy scheme

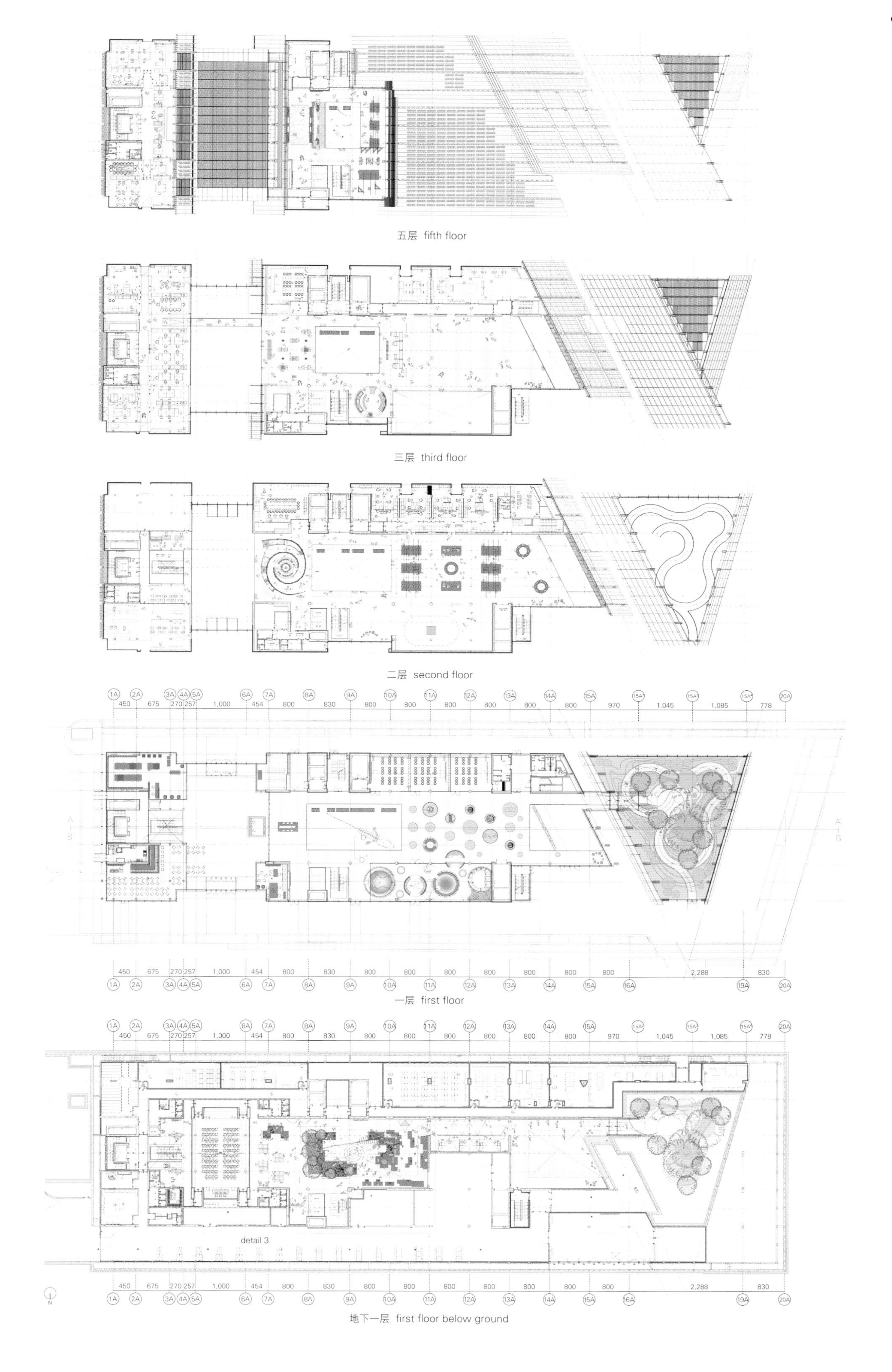

五层 fifth floor

三层 third floor

二层 second floor

一层 first floor

地下一层 first floor below ground

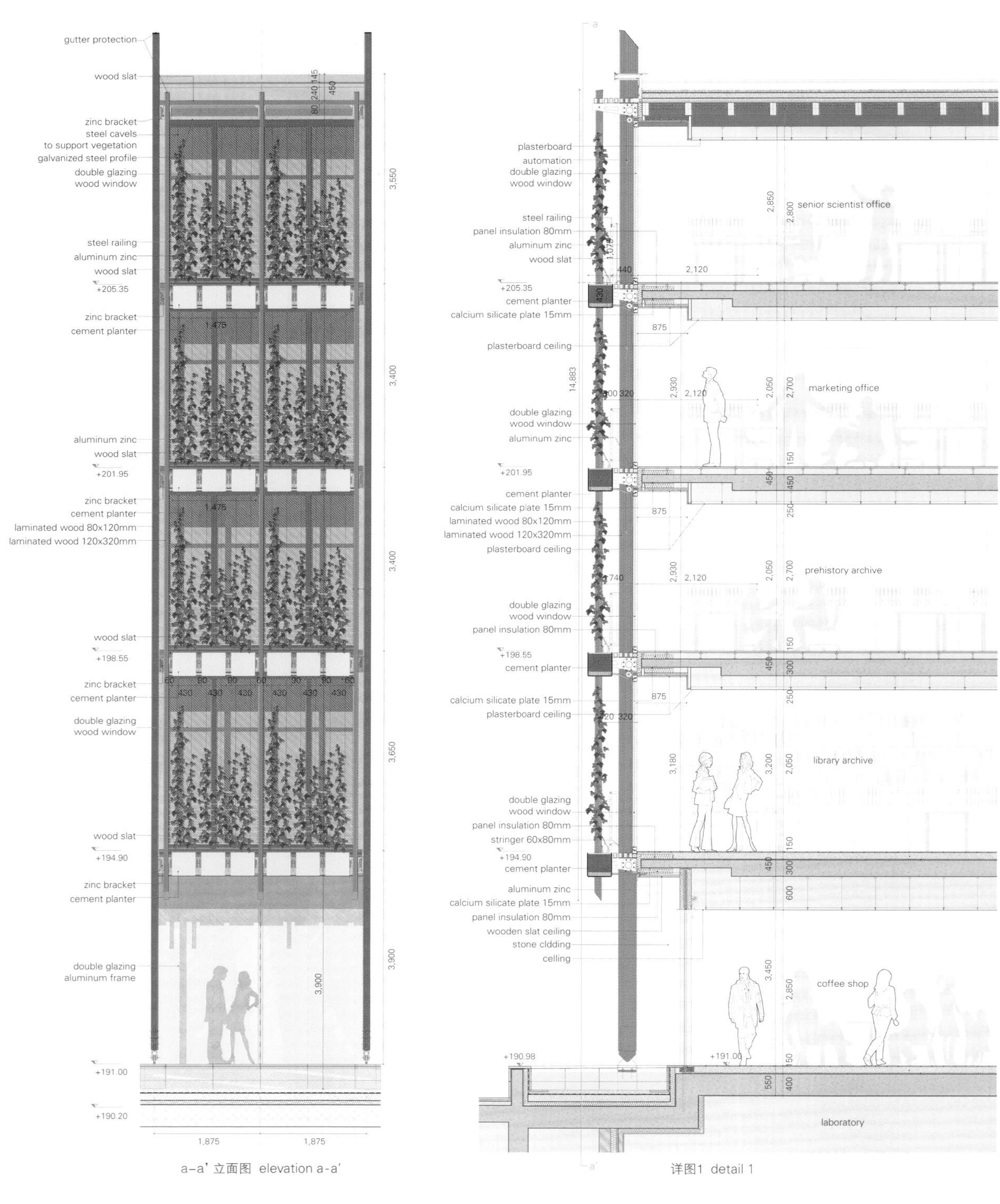

a–a' 立面图 elevation a-a'

详图1 detail 1

thanks to the clearer definition of the specific architectural elements that make up the rest of the district itself, above all in terms of its tertiary, residential and commercial functions.
The building is made up of a sequence of spaces and volumes, solids and voids resting or seemingly floating upon a large body of water, thus multiplying the effects and vibrations of light and shade. The entire structure is held together at the top by its large roof layers, which are in complete harmony with its forms, thus rendering them recognizable even from the outside. Starting from the east, the first structure houses functions which are not available to the public, such as administrative and research offices, scientific laboratories and ancillary spaces for on-site staff.
Next, we find the lobby. It is aligned with the main axis of the district and traverses the entire depth of the building towards the north, overlooking the park area outside Palazzo delle Albere.
The scientific themes of the mountain and the glacier are subsequently dealt with through a series of exhibition spaces, which gradually rise up from the basement level and nearly "break through" the roof, thus creating an observation point immersed within the environment, from which a true "simulation" of the real experience can be enjoyed. This experience is highlighted by ample exhibition spaces on two or three levels, with ceilings high enough to welcome extremely large sets and backdrops.
The building's shape or "rain forest" function also serves to define its interior space and functionality. In fact, the building represents a large tropical greenhouse which, during certain periods of the year, is even capable of establishing a functional relationship with the specific exhibition stands even outdoors, in which water, lighting and greenery often play a key role in defining the visitor's natural surroundings.
The educational and laboratory services for the public are offered in a series of aboveground structures located alongside the exhibition areas, thus promoting interactive experiences for each individual subject matter.

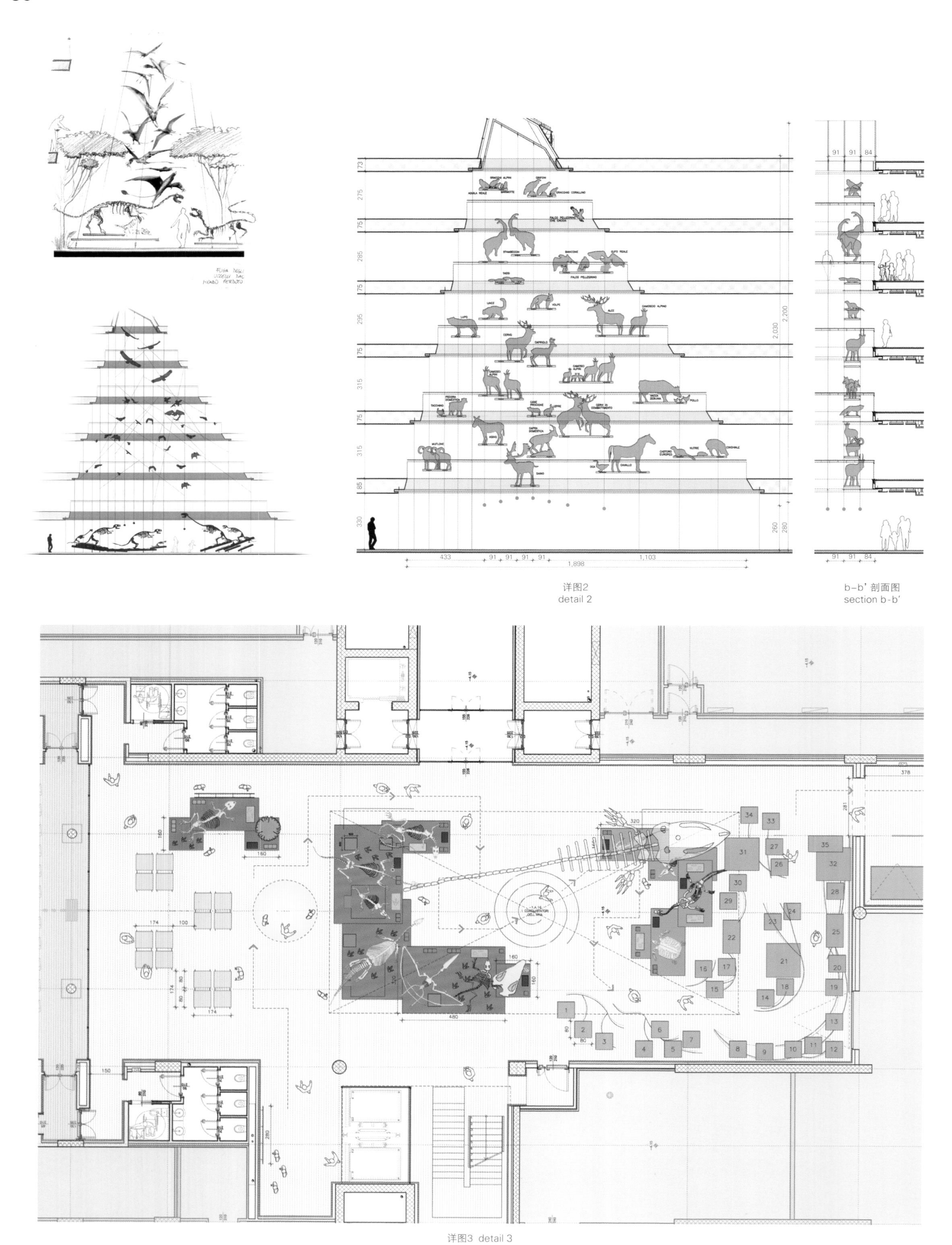

详图2
detail 2

b–b' 剖面图
section b-b'

详图3 detail 3

项目名称：Muse and "Le Albere" Area
地点：Trento, Italy
建筑师：Renzo Piano Building Workshop
合伙人和副主管：S. Scarabicchi, D. Vespier
项目团队：A. Bonenberg, T. Degryse, E. Donadel, V. Grassi, F. Kaufmann, G. Longoni, M. Menardo, M. Orlandi, P. Pelanda, D. Piano, S. Polotti , S. Russo, L. Soprani, G. Traverso, D. Trovato, C. Zaccaria, C. Araya, O. Gonzales Martinez, Y. Kabasawa, S. Picariello, S. Rota, H. Tanabe / CAD operator_S. D'Atri / models_F. Cappellini, A. Malgeri, A. Marazzi, S. Rossi, F. Terranova
结构工程师：Favero & Milan / 建筑服务：Manens Intertecnica / 能源工程师：Associazione PAEA
音响工程师：Müller BBM / 成本顾问：Dia Servizi / 水文地质研究：M.Vuillermin
道路与辅助设施设计：A.I.A. Engineering / 污水网设计：Ingegneri Consulenti Associati
防火设备成本顾问：GAE Engineering / 景观设计：Atelier Corajoud-Salliot-Taborda, E.Skabar
造价与技术顾问：Tekne / 项目协调：Twice/lure / 甲方：Castello SGR S.p.A.
用地面积：116,331m² / 总建筑面积：20,000m² (with underground surface)
有效楼层面积：97,460m² / 地下表面积：113,000m²
造价：EUR 240 million / 设计时间：2002—2009 / 施工时间：2009—2013
摄影师：
©Paolo Pelanda(courtesy of the architect) - p.80
©Stefano Goldberg - Publifoto(courtesy of the architect) - p.83
©Enrico Cano(courtesy of the architect) - p.76~77, p.78~79, p.84, p.87, p.88~89, p.91
©Alessandro Gadotti(courtesy of the architect) - p.82, p.85

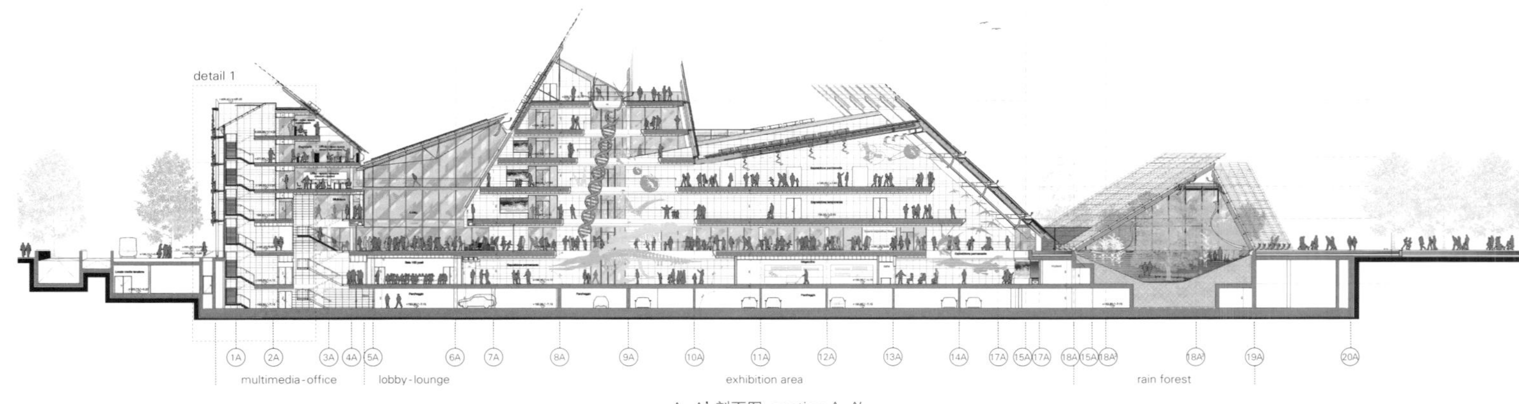

A-A' 剖面图 section A-A'

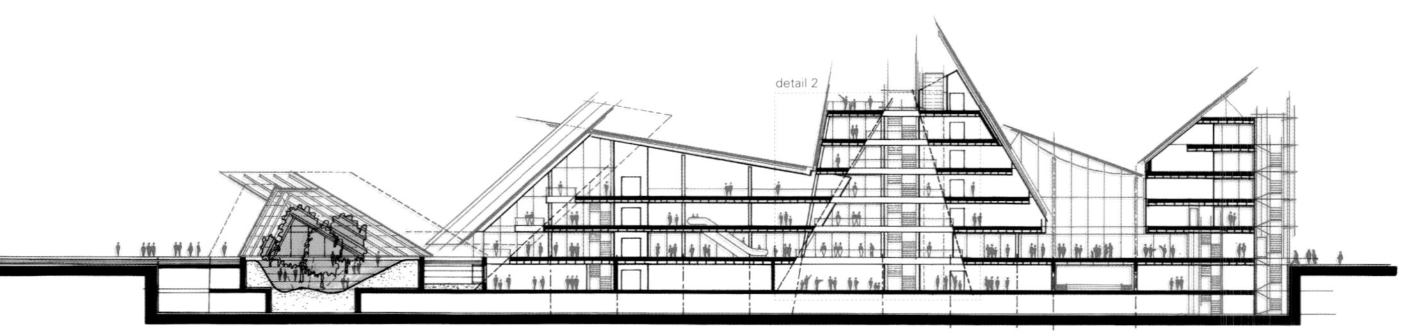

B-B' 剖面图 section B-B'

七抹绿荫 Seven Shades of Green

社会越来越崇尚见多识广和激情，这种趋势使我们开始积极寻求更加可持续的未来。在这方面，建筑师通过创造更高效和环保的建筑，在探索潜在的道路上扮演着至关重要的角色。实际上，建筑学依赖创新的解决方案，将过去的知识与新技术结合，来预测新一代尖端的可持续建筑。可持续发展的新浪潮略过了20世纪90年代的以建筑的"绿色"生产为特色的实验主义，而是在建筑语言和能量消耗的目标方面，似乎已经走向了成熟。这很可能是在设计过程中与一些参考相结合的结果：高效能源不再是人们关注的唯一且简单的问题；相反，环境作为一个整体包含了文化层面，成为人们的指导原则。

There is a growing trend of a well-informed and passionate part of society that is driving us all towards a more sustainable future. In this regard, architects play an important role in exploring potential pathways, through the creation of more efficient and environmentally sound buildings. In fact, architecture is seeing a new generation of cutting-edge sustainable buildings, which rely on innovative solutions combining new technologies with knowledge from the past. Having moved past the experimentalism that characterised the architectural "green" production of the 1990s, the new wave of sustainability appears to have come to maturity in terms of language, as well as energy consumption objectives. This is a likely result of the combination of several references in the design process: energy efficiency is no longer the sole and simple center of attention; instead, the environment as a whole – including cultural dimensions – has become the guiding principle.

如今，绿色和可持续性似乎已经成为解读建筑实践和生产变化的关键词。根据不同的建筑使用背景，作为一个形容词，"可持续的"有多重含义。

纵观近期已经被归类为可持续的建筑项目，我们可以通过不同的方法去感知可持续性：一些建筑师采取高科技的解决方案，而其他建筑师却恰恰相反，使用基本的不使用能源的设备；在一些案例中，乡土建筑被重新采用，而在其他过往的建筑里，文化要素支配着整个进程。通过一系列的方法，海伦·本尼茨、安东尼·拉德福德、泰瑞·威廉逊[1]总结出三条基本原则：自然，暗示了对生态系统的关注；文化，考虑了地区的多样性；技术，依靠设备的使用性。建在加拿大魁北克省的葛兰素史克生物制品行政大楼，完美地被东西轴线所指引。其北面的立面与南面的立面完全相反，但是都在建筑的可持续性方面扮演着重要的角色。事实上，北立面设计在内部通过实体墙来聚集热能，同时一面大型全玻璃双层立面使建筑面向南方。这个设计在冬季储存热量，而在夏季，空气在室内外之间的空腔内流动，来减少热量。向南的玻璃窗使内部空间在太阳光线中一点点被淹没，以此突出优雅的木质结构。木材的选择并非偶然，实际上它可以帮助整座建筑保持碳平衡。

例如，Lake Flato建筑师事务所和RSP建筑师事务所建造的亚利桑那州立大学的理工学院教学楼证实了在设计过程中环境策略成为了理论原则和工具。曾经面积为56 656m^{2}的混凝土和沥青才能修建出一个空军基地的地方，现在，五栋大楼就已经形成了亚利桑那州立大学校园的主干。从远处看，它们被看作是强大的立体式组合，但是它的复杂性，以及某部分材料的稀少，逐渐地在近距离内暴露出来。这种转变是建筑意向的关键：一栋建筑能融入周围景观，同时还能创造一系列迷人的社交互动空间。这个双重策略的产生是了不起的公投结果。沙漠景观塑造了建筑，设定了规模、色彩和风情。在校园里面，它们属于教学区域之间的空间，并且穿过教学区，成为最迷人的建筑。这处空间由步行通道和本土花园组成，这些花园将围绕主体建筑的庭院连接起来。步行区的作用

Nowadays green and sustainable seem to be key words for reading changes in architectural practice and in its production. "Sustainable" is an adjective that groups several shades of meaning according to the architectural context used.
Looking into recent projects that have been classified as sustainable, it is possible to underline different approaches to achieve it: some architects employ high-tech solutions, and others go in the opposite direction using basic passive devices; in some cases the vernacular construction is rediscovered, while in other experiences the cultural component becomes dominant in the process. Trying to group these approaches, there appear to be three common principles, as noted by Helen Bennetts, Antony Radford, Terry Williamson[1]: nature, which implies attention towards the ecosystems, culture, which takes into consideration the diversities among places, and technology, which relies on the use of devices. GlaxoSmithKline Biologicals Administrative Building in Quebec, Canada, is perfectly orientated on an East-West axis. Its northern facade is diametrically opposed to the southern, and both play a key role in the building's sustainability. Indeed, the northern elevation is designed to conserve heat inside the building with a solid wall, whilst a big, fully-glazed double-skinned facade characterises the building towards the south. This element also traps the heat in winter whereas air flow through its intermediate cavity reduces the heat in summer. The south-facing glazing allows the internal space to be gently flooded by sun rays, exalting in this way the graceful timber structure. The choice of the timber is obviously not accidental, in fact it is a material that makes the structure carbon neutral.
An example of the way in which environmental strategies become theoretic principles and tools for the design process is the ASU Polytechnic Academic Buildings by Lake Flato Architects and RSP Architects. Where once 14 acres of concrete and asphalt constituted an airbase, now, five buildings form the spine of the Arizona State University campus. The construction is seen as a powerful stereometric composition from afar, however its complexity and, in some parts, the rarefying of its materials are revealed at a closer range. This transition is the key for the architectural intent: a building able to merge into the landscape and at the same time, create a series of captivating spaces for social interaction. The result of this dual strategy is a remarkable civic outcome. The desert landscape shapes the buildings, setting their scale, colours and disposition. In the campus it is the space in between the teaching areas and across them emerges as the most captivating. This space consists of pedestrian paths and native gardens that link the court-

照片提供：©Herzog & de Meuron (Michael Bär)

里恩天然游泳池，里恩，瑞士
Naturbad Riehen, Natural Swimming Pool, Riehen, Switzerland

相当复杂：它是人性化的设置，由于庭院的存在，它保护游客不受周期性季风的伤害，且提供对环境敏感的社交空间，以尽可能地提高可视性和社区意识，使日光尽可能的多。而且，户外通道使内部的流通空间最小化，因此显著降低了能量配额，否则就需要使用空调。

有些时候，可持续性的文化元素都体现在了某一地域的设计结果方面。

这一点可以在Branch建筑师工作室设计的位于澳大利亚的圣莫尼卡学院的翻新和扩建中有所体现。该项目涉及对现有的学校图书馆的改造，修建新的阅读休息室，扩建平台，平台能赋予这处游乐环境以活力，而此地的若干处空间——门槛、花园门厅、多元化服务台、西班牙式阶梯，以及树屋式的阅读休息室——都通过斜面屋顶所界定的动态空间来相互连接。该建筑通过包含实际和隐喻元素的连接策略来与澳大利亚的自然环境完美融合。这种内外之间的相互作用和结合最为人称道之处是在平台上可以俯瞰国家公园：户外体验可以作为学生们的文化课程。同样，无论是在内部还是在长长的庭院艺术品（抽象的波奴鲁鲁国家公园鸟瞰图）方面，使用强烈的色彩都是在向澳大利亚内陆（指澳大利亚等偏僻而人口稀少的地方）致以敬意。

然而，在全球范围内一些可持续经历都没有单一地通过地点，在文化或者历史方面展示出来。这是因为"绿色"法则不适用于某些特定地区，而是仅适用于某些气候环境。

例如，位于科瑞亚诺的生态性综合建筑由意大利的Triarch工作室建造，建筑师努力寻求各种各样的生态学原则，来创造一个充满技术知识的综合体。这栋建筑可以用来举办同可持续性和生态相关的展览、活动和研讨会，成为一个向环境表达敬意、为每一位建筑师提供参考的活生生的典范。其设计策略就是根据当地的生态足迹，就地取用原材料（对当地材料的估计总量及明确的系统中将会产生的废弃估计量进行了流动分析）、能量消耗最小化、进行热保温、广泛使用的自动化和运用可再生能源。光照和热量摄入的控制，都可以通过建筑体量的巧妙设计来实

yards that are wrapped around the main buildings. The function of the pedestrian area is extremely complex: it generates a human scale; it protects the visitors from the seasonal monsoons thanks to its courtyards; it provides environmentally sensitive social spaces that maximise visibility, daylight, and the sense of community. Furthermore, the outdoor paths minimise the internal circulation space, thereby significantly reducing the energy quota that would otherwise be required for air-conditioning.

Sometimes the cultural components of sustainability are embodied in designs that are clearly the result of certain geographical territories.

This can be seen, for example, in the renovation and extension of the St Monica's College in Australia by Branch Studio Architects. The project involved the alteration of the existing school library, with the creation of a new reading lounge and a deck extension which gave life to a playful environment in which several spaces – the entry threshold, the garden foyer, the multi-desk, the Spanish steps, and the tree-top reading lounge – are interconnected in a highly dynamic space defined by sloping roofs. The building engages with the Australian natural environment through an articulate strategy comprised of factual and metaphorical elements. The interaction and fusion between interior and exterior, sublimated into the deck that overlooks the national park, are the most outstanding: the outdoor experience as a cultural lesson for students. Similarly, the intense palette of colours used pays homage to the Australian outback, both in the interiors and in the long courtyard artwork – an abstraction of an aerial view of the Bungle Bungle National Park.

Nevertheless, certain sustainable experiences around the globe are not univocally connoted by the location in terms of culture or history. This is because the "green" principles applied are not specific to a certain region, but rather adapted only to the climatic conditions.

For example, the Ecoarea Complex by Triarch Studio in Coriano, Italy seeks to apply a wide range of ecological principles in its architecture, creating a complex synthesis of technological knowledge. The building, designed to house exhibitions, events and seminars related to sustainability and ecology, becomes a living reference for each architect that shows respect toward the environment. The design strategy involved the selection of materials according to their ecological footprint (the flow analyses to estimate the total material and waste generated in a well-defined system) and minimal energy consumption, implicating thermal insulation, extensive automation, and the use of renewable en-

照片提供：© Branch Studio Architects (Nils Koenning)

帕梅拉·科因图书馆，维多利亚，澳大利亚
Pamela Coyne Library, Victoria, Austraila

现。从外面看，建筑犹如一个切开的白色体块，部分得以显示，以展示内部的部分复杂布局。内部空间是由一个连续的表面形成，表面经过折叠，呈螺旋状，围绕在一个上空体量的周围，形成地面和连接的坡道。采用这种方式，建筑形成了一系列的空间，这些空间能够最优化地利用直射日光，并且根据它们的功能，确保自然光贯穿整座建筑。

由阿克雅协会和Favero&Milan工程公司联手为上海交通大学闵行校区所设计的绿色能源实验室，既大量使用了绿色技术概念，又与本土文化相结合。建筑师们构思出这个研究中心和实验室是为了分析和传播施工和居住方面应用的低碳排放技术，作为其规划的赘述。实际上，建筑处于被树木包围的隔离区，其体量紧凑，并围绕一处上空空间布局，这处空间的功能如同热风筒，能够优化能源效率。而这处空间在冬天成为热量的聚集地，夏天却被当作排气道，将内部产生的热空气吸出，并通过顶部的大型可开启的天窗将它们排出。同样，外部精修的双层立面也有双重作用。从活力的角度分析，陶质的百叶窗使建筑的外围护结构镶嵌有花纹，为内部的工作空间调节光线，而内层的玻璃隔间能够防水和保温。从视觉的角度分析，立面装饰和传统的中国木屏风有明显的异曲同工之处，整个图案利用精致的中文文字取代了常见的几何构图。

这些例子已经列举了自然、文化和技术方面的多种可持续方法。上文已经列举过的建筑有不同的规模和象征意义，然而每一个都显示出与环境建立密切对话的承诺。希望可以朝着这个方向努力，接下来在绿色议程中，"可持续"这个词已不再需要，因为它的意义和它的原则已经完全融入到每一个建筑项目中。

ergy sources. The control of the light, and therefore of thermal intake, is achieved through skillful design of the building volume. From outside, it appears as a white mass that has been carved, then partially screened to reveal a portion of the complex internal layout. The interior space is created by a continuous surface folded into a spiral around a void to create the floors and connecting ramps. In this way it generates a series of spaces which make the best use of direct sunlight, according to their function, and assure natural lighting throughout the building.

A building that largely uses both green technical concepts and creates a certain bond with the local culture, is the Green Energy Laboratory, designed by Archea Associati in collaboration with the engineering firm Favero & Milan for the Minhang Campus of the Jiao Tong University, Shanghai. The architects conceived this research center and laboratory for the analysis and diffusion of low carbon emission technologies in the construction and housing sector, as a sort of tautology of its program. Indeed, isolated on a site surrounded by trees, the building has a compact volume and is organised around a void that functions as a thermal chimney, to optimise energy efficiency. The void acts as an accumulator of heat in winter, whereas in summer as a flue, which aspirates the hot air produced in the interior and releases it through the big openable skylight at the top. In a similar way, the external, elaborate double-skinned facade has a dual function. From a energetic point of view the earthenware shutters that tessellate the building envelope regulate the light in the working spaces inside, while an internal layer of glazed cells provides waterproofing and insulation. If analysed from a visual point of view, the facade ornament relates clearly to traditional Chinese timber screens, in which patterns derived from elaborating Chinese characters replace the usual geometric compositions.

These examples illustrate a handful of the multiple approaches to sustainability that revolve around nature, culture and technology . The buildings that have been discussed have different scales and various figurative results, however each of them demonstrates a commitment toward establishing a tight dialogue with the environment. It should be hoped that in moving in this direction, the next step in the green agenda would entail that the word "sustainable" is no longer needed, for its meaning along with its principles becoming integrated into every architectural project. Simone Corda

1. Helen Bennetts, Antony Radford, Terry Williamson, *Understanding Sustainable Architecture*, London: Spon Press, 2003, p.25.

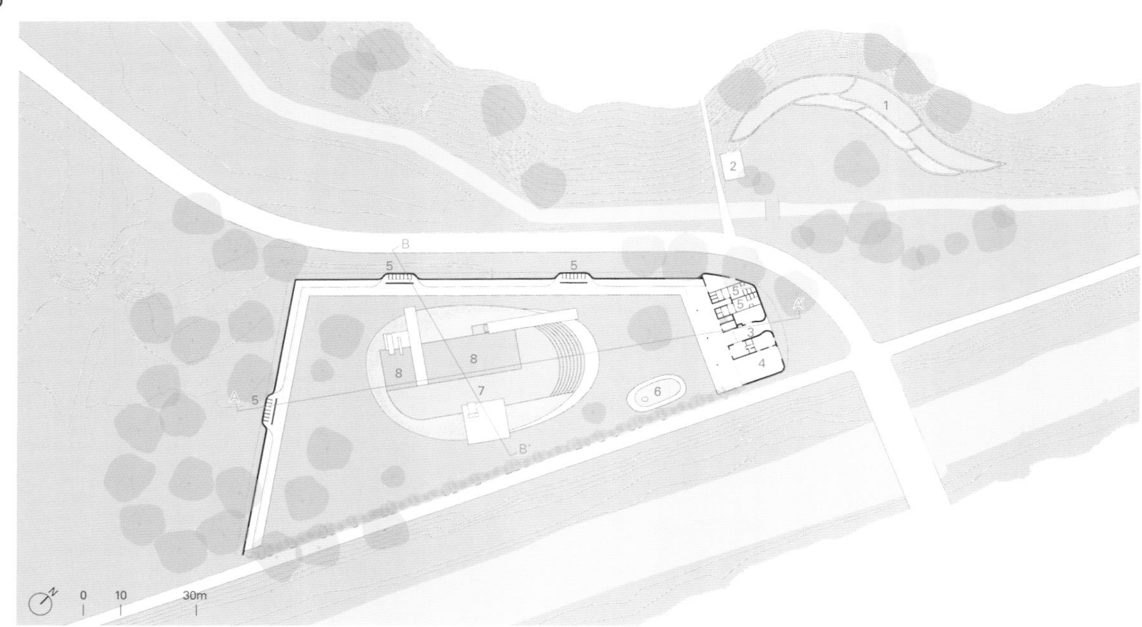

1 过滤系统
2 发电站
3 入口
4 咖啡室
5 更衣室/卫生间/淋浴室
6 儿童泳池
7 非游泳区
8 游泳区

1. filter system
2. power station
3. entrance
4. cafe
5. changing room/toilet/shower
6. children's pool
7. non-swimming area
8. swimming area

里恩天然游泳池

Herzog & de Meuron

瑞士的里恩市毗邻巴塞尔，位于与莱茵河交汇的、正缓慢拓展的威斯河峡谷。数十年来，当地人殷切地渴望修建一个新的公共游泳池来取代已经逐渐被淘汰的堤岸旁的浴室，不过一次又一次的努力都化作了泡影。

赫尔佐格·德梅隆在1979年赢得了一场设计比赛后，在接下来的几年内若干个项目都没有建成。他开始重新考虑修建一个新的公共浴场设施。几年间不断改变的透视图，促使他放弃使用机械和化学水处理系统的传统泳池理念，进而建造一个接近具有生物进化功能的自然环境的泳池。这项提议得到莱茵地区民众的激烈讨论，并进行了民意表决。从标准的几何形游泳池转变成一个没有技术系统和机械间的、直接种植过滤性植物系统的天然洗浴湖。这个理论缔造了在"巴蒂"修建天然泳池的概念，巴蒂是巴塞尔传统的木质浴室（位于莱茵河畔），将活跃的气氛和永恒的外观很好地结合起来。

场地被两侧的木质墙体所围合：北面邻街，西面紧挨私人产业，南面边界则靠河，换句话说，是开放的，只被绿篱围绕。在东边，设施建筑里融入了一个木栅栏，还设有入口和辅助设施，同时西面和北面的墙体提供了200m长的、可以遮阳的日光浴躺椅。然而，无论从设施的哪个部分来说，人们的注意力都会集中在场地中心的浴池。作为浴室的非机械化核心，生物水处理池嵌在道路另一侧倾斜的景观内。建筑与这里所提供的各种休闲设施一起，共同形成了一处全年面向市民开放的休闲区。从生态清洁能力的角度来看，浴场每天能够接受2000位游客的到访。

Naturbad Riehen, Natural Swimming Pool

The Swiss municipality of Riehen, bordering the city of Basel, lies in the gently widening valley of the River Wiese, near to its confluence with the Rhine. For decades, the local population has yearned for a new public swimming pool to replace the obsolescent baths by the riverbank, with various attempts having failed. After winning a design competition in 1979 and several unrealized projects in the following years, Herzog & de Meuron again started to ponder the options for a new bathing facility. The changed perspectives brought by the intervening years prompted the idea of abandoning the conventional pool concept with its mechanical and chemical water treatment systems in favor of a pool closer to a natural condition with biological filtration. This approach

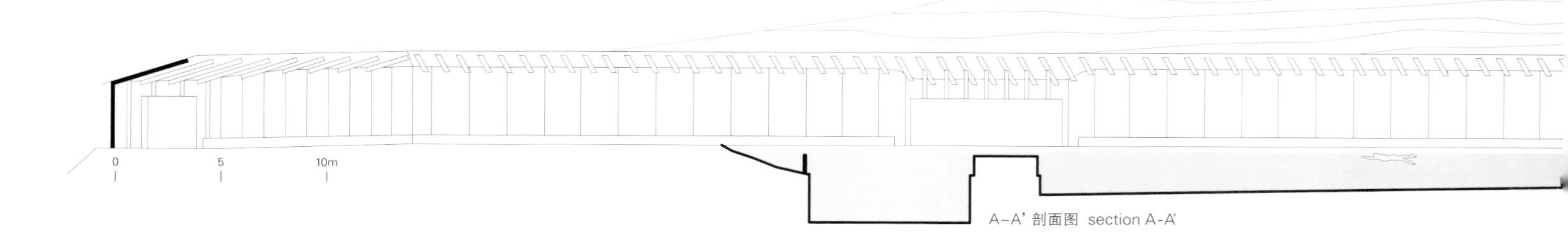

B-B' 剖面图 section B-B'

A-A' 剖面图 section A-A'

was publicly discussed by the citizens of Riehen and officially approved by a municipal vote. The standard geometric swimming pool transforms into a bathing lake where the technical systems and machine rooms vanish, to be substituted by planted filtering cascades. This concept led to the notion of modeling the natural pool in the local "Badi", Basel's traditional wooden Rhine-side baths, which combine a lively atmosphere with a timeless appearance.

The site is screened on two sides by an enclosing timber wall: on the north towards the road and to the west from adjoining private properties. The southern perimeter facing the river, on the other hand, is open, bounded only by a green hedge. On the eastern front, a timber fence merges into the amenities building, which incorporates the entrance and supporting facilities, while the wall along the northern and western boundaries offers a 200 m long sheltered solarium with recliners. Yet, from all parts of the facility, the attention is focused on the bathing pond at the center of the site. The biological water treatment basins – the non-mechanical "heart" of the baths – are embedded in the sloping landscape on the opposite side of the road. Together with various leisure facilities provided here, they form a recreational area open all year around to the municipal population. In terms of ecological cleaning capacity, the baths are designed to accommodate 2 000 bathers per day.

项目名称：Natural Swimming Pool / 地点：Weilstrasse, Riehen, Switzerland
建筑师：Herzog & de Meuron
项目合伙人：Jacques Herzog, Pierre de Meuron, Wolfgang Hardt(until 2011.4.30)
项目团队：Associate, Project manager _ Michael Bär / Project manager _
Harald Schmidt, Sarah Righetti / Jeanne Autran, Nathalie Birkhäuser, Nils Büchel, Thomas Cardew, Judit Chapallaz-Laszlo, Dorothee Dietz, Guillaume Henry, Guy Nahum, Uta Schrameyer,Tobias Josef Fritzenwenger, Benno Lincke, Miguel Palencia Olavarrieta
合作建筑师/总规划师/造价顾问：Rapp Arcoplan AG
电气工程师：Eplan
暖通空调工程师：Stokar + Partner AG
景观设计师：Fahrni und Breitenfeld, Wasserwerkstatt
水暖工程师：Locher Schwittay Gebäudetechnik GmbH
结构工程师：Ulmann & Kunz Bauingenieur AG, Pirmin Jung
其他专家：Wasserwerkstatt Planungsbüro für Badegewässer
木工：Pirmin Jung / 建筑物理：Ehrsam und Partner
土木工程师：Gemeinde Riehen / 几何学家：Jermann Ingenieure & Geometer AG
土工技术顾问：Dr. von Moos AG
甲方：Gemeindeverwaltung Riehen, Riehen, Switzerland
用地面积：15,243m² / 有效楼层面积：324m²
材料：facade _ wood construction, larch / roof _ coated aluminium sheeting / stone floor _ lissone quartzite
设计时间：2007.8—2012.12 / 施工时间：2013.4 / 竣工时间：2014.6
摄影师：
Courtesy of the architect-p.100~101
©Leonardo Finotti-p.96~97, p.98~99, p.102~103

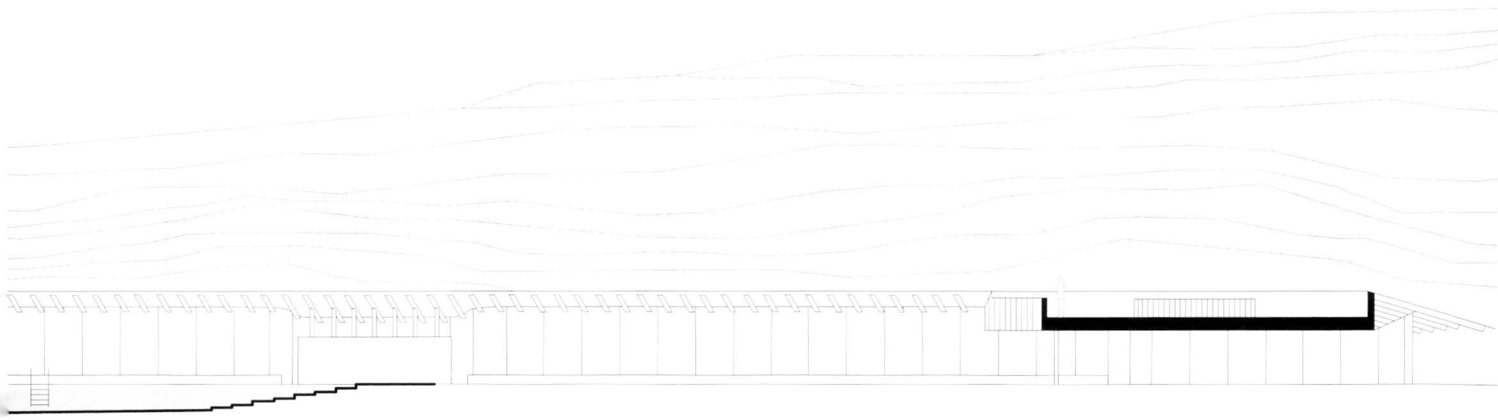

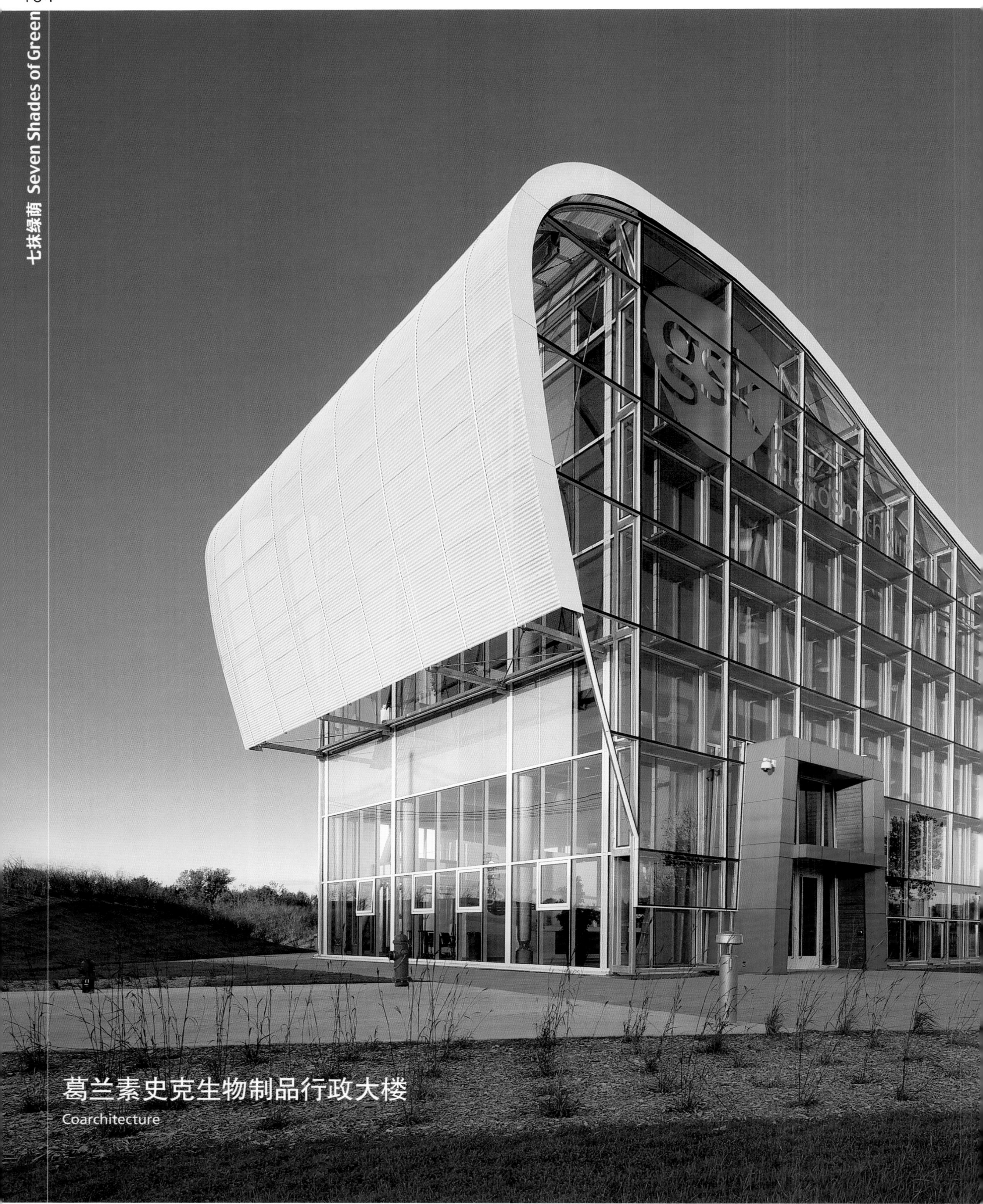

葛兰素史克生物制品行政大楼

Coarchitecture

项目名称：The Administrative Building of GlaxoSmithKline
地点：Quebec City, Quebec, Canada
建筑师：Coarchitecture
结构工程师：SDK et associés
机械工程师：Pageau Morel et associés inc.
景观建筑师：François Courville architecte paysagiste
土木工程师：Marchand Houle
施工管理：Verreault
甲方：GlaxoSmithKline
功能：Administrative building for a biothchnology company
用地面积：37,000m²
竣工时间：2011
摄影师：©Stéphane Groleau (courtesy of the architect)

魁北克城的葛兰素史克建筑的设计目的是为了支持流感疫苗工厂的管理和维护活动。2009年，这个国际大企业邀请四家建筑公司参与竞标。项目书要求设计能效高且极富特色的建筑典范。Coarchitecture建筑事务所竞标成功，因为其设计以价值创新为基础，以可持续发展目标为宗旨。赋予这栋建筑独特性的推动力是通过整合所有相应的系统，来展示其木质结构。建筑场地泥土质量低劣，几乎没有植被覆盖。其场地规划和景观设计是尽量让它还原到自然状态。促进生物多样性，完成地下水的补给，以确保再引进的本地植物的生长，建成后的景观不仅有利于提高内部空间的质量，而且还赋予住户活力，所有的一切都与木材作为天然的材料息息相关。

同样，带有有机轮廓的建筑已被放置在这里与环境共生，就像动物巧妙地适应影响它生存的周遭环境一样。它面向南方，一个遮阳系统融入到全玻璃的双层外围护结构中，保护建筑免受太阳热能的侵扰，同时还十分美观。外围护结构的东侧和西侧带有透明的玻璃嵌板，而在北侧，它还能为寒冷的冬季节约能源。这些技术还可以通过充足的自然光线来提供舒适的室内环境，据住户称，还起到了鼓舞的作用。

采用全玻璃双层立面的一个重要的目的就是把建筑内的木质结构向街道和高速公路上的人们展示。同样，木材对建筑的美观方面做出了重大贡献。Coarchitecture建筑事务所认为这是最大限度地提高木材在高端建筑中使用的正确方式，以避免维护问题，并且确保长期的低维护费用。

为了通过全玻璃立面来获得高效能量，且提高员工的生活质量，生物气候设计原则必须得到尊重。主要的工作区域在北面，以保护员工不受刺眼的强光和太阳辐射波段的伤害，提供全天畅通无阻的视野。辐射供暖系统为室内供暖，同时在低楼层的木质平台起到隔音的作用。冷梁和漫射光线的吸音板组成一个自定义系统，在不影响声音质量的同时还能体现木质结构的丰富性。

南立面一侧是所谓的"互动空间"和休息室，用来鼓励住户小憩或者进行社交活动。鉴于其用途，这里每处空间的温度都可以通过自然通风来调节，或者在短暂的冬季通过直接增加太阳辐射的方式来形成丰富而充满活力的照明。

居民们表达出的非常一致的满意度，已经转化为创造力、生产力的提高以及公司效益的增加。因此，这个项目成功地显示出行政管理部门在提高城市景观绿化质量方面，尤其在实现可持续发展方面，有着巨大的潜力。

GlaxoSmithKline Biologicals Administrative Building

The GlaxoSmithKline Building in Quebec City was built to support the administrative and maintenance activities of the influenza vaccines plant. In 2009, the international company invited four architectural firms to participate in a competition. The brief was to design a highly distinctive building that would also be exemplary in energy efficiency. Coarchitecture's proposal won based on value innovation aimed at sustainable development targets. The main driver that gave the building its uniqueness was to showcase a wood structure by integrating all the systems accordingly. The building site had poor soil quality and was practically void of vegetation. The site planning and landscaping approach were to

南侧双表皮
1 抽吸通风
2 高性能窗框
3 通道/遮阳板
4 新鲜空气进气口

空气处理
5 空气进气口
6 空气处理单元
7 冷梁
8 回风口

9 送风区
10 夏季：
A-自然通风
B-机械通风
冬季：
机械通风
11 废气口

雨水管理
12 屋顶排水管
13 氯化处理

供暖和制冷
14 地热井
15 热泵
16 热交换器
17 辐射采暖地面
18 冷梁供应

照明
19 间接的人工照明
20 低碳双层玻璃（夏季）
21 百叶窗（冬季）

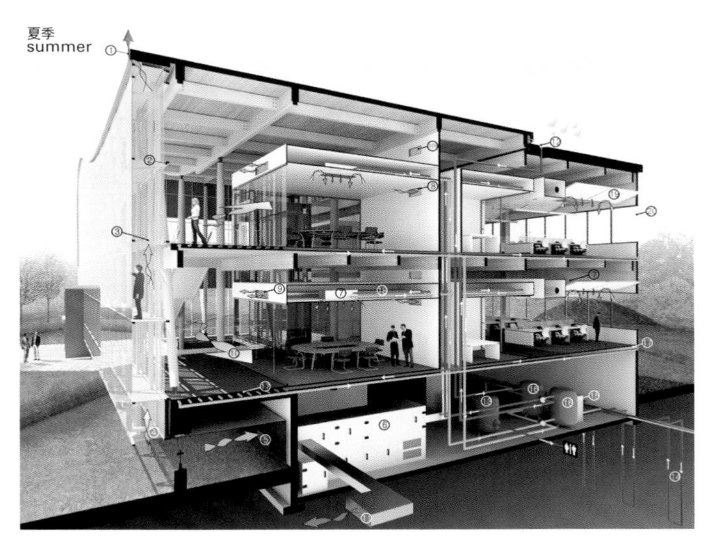

south double skin
1. chimney effect ventilation
2. high-performance mullions
3. walkway / brise-soleil
4. fresh air intake

air management
5. air intake
6. air handling unit
7. chilled beam
8. air return

9. air supply
10. summer:
A-natural ventilation
B-mechanical ventilation
winter:
mechanical ventilation
11. air exhaust

rainwater management
12. roof drain
13. chlorine treatment

heating and cooling
14. geothermal wells
15. heat pump
16. heat exchanger
17. radiant floors
18. chilled beam supply

lighting
19. indirect artificial lighting
20. low-e double glass (summer)
21. blinds (winter)

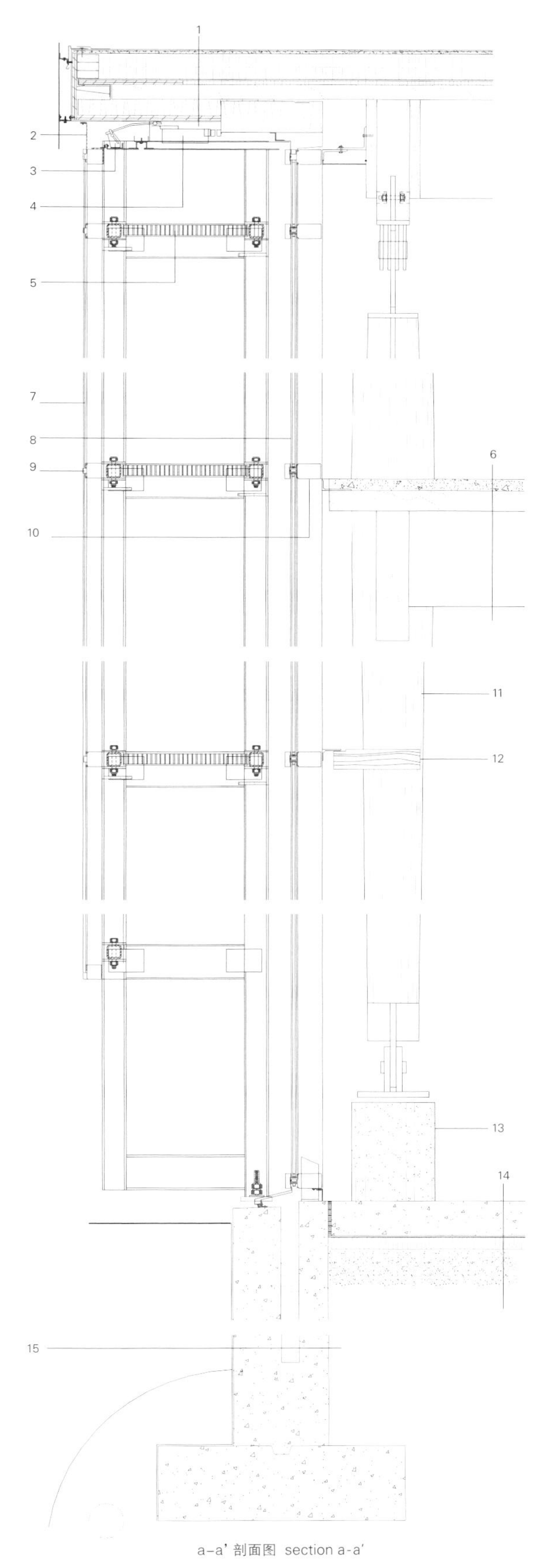

a–a' 剖面图 section a-a'

1. roof assembly
2. birdscreen
3. vent damper
4. motor for damper control
5. galvanized steel walkway/ brise-soleil
6. floor assembly
7. single-pane glass
8. double-pane sealed glass unit
9. double-skin aluminum
10. aluminum curtain wall
11. wood column
12. wood grid
13. concrete base
14. ground-floor assembly
15. foundation wall

restore it to an indigenous natural state. Promoting biodiversity and the replenishment of groundwater to ensure the growth of the native plants that were reintroduced, the finished landscape contributes not only to the quality of the interior spaces but to the reenergizing of the occupants as well, all in coherence with the presence of wood as a natural material.

Similarly, the building with its organic silhouette has been placed here in symbiosis with the climate, much like an animal ingeniously adapted to its environment as a matter of survival. It faces the south, protected aesthetically from solar heat gains by a system of sunshade integrated into a fully-glazed, double-skinned envelope with translucent metal panels on the east and west sides. On the north side, the building envelope is designed to conserve energy in the cold winter. These techniques have been employed to provide comfortable, indoor environment lit abundantly by natural light which, as their occupants have reported, has an invigorating effect.

An important objective with the fully glazed double skin facade was to expose the wood structure to the street and highway views. Again, the wood brings a highly remarkable contribution to the building's aesthetics. Coarchitecture believes this is the right way to maximize wood use in high end buildings to avoid maintenance issues and insure long term performance at low cost.

To achieve high energy efficiency as well as improved quality of life for the workers with a fully glazed facade, bioclimatic design principles need to be respected. The main work areas were placed on the north side to shield them from glare and fluctuations in solar radiation, and to provide an unobstructed outside view throughout the day. A radiant heating system provides warmth while at the same time, soundproofing the wooden decking on the lower level. Chilled beams and acoustic panels that diffuse the light make up a customized system that allows the richness of the structural wood to be seen without compromising the quality of the acoustics.

Lining the south facade are the so-called "interactive spaces" and lounges where the occupants are encouraged to take their break and socialize in a comfortable and relaxing environment. Depending on their usage, the temperature in each of these spaces can be controlled through natural ventilation or by increasing direct solar radiation to create a rich, energizing luminosity which acts as light therapy in the short winter days.

The unanimous satisfaction expressed by the occupants of the building translates into increases in creativity, productivity and benefits for the company. The success of this project, therefore, has the potential to change attitudes in a way that the administrative sector can better contribute to the quality of the urban landscape, and especially, towards sustainable development.

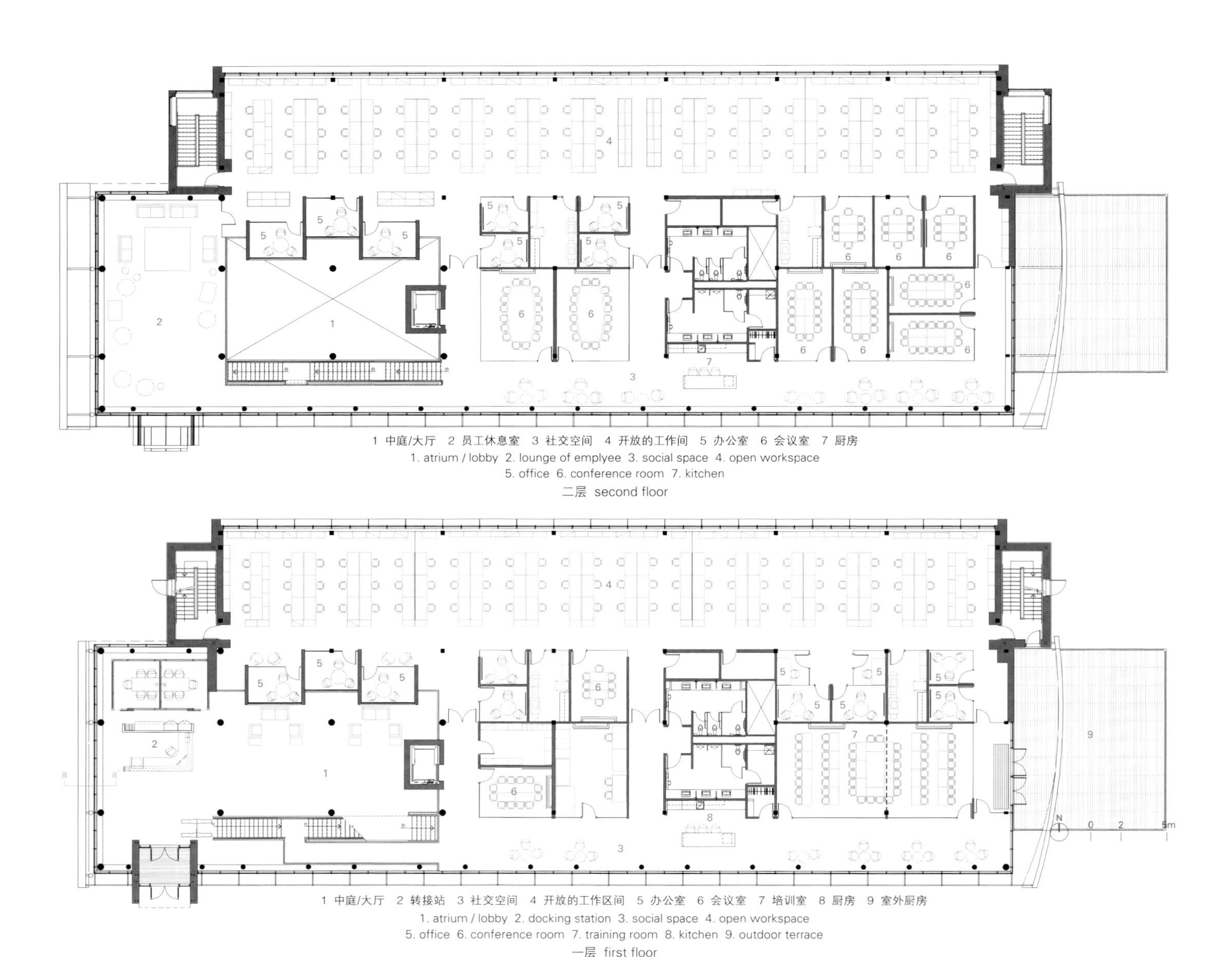

1 中庭/大厅　2 员工休息室　3 社交空间　4 开放的工作间　5 办公室　6 会议室　7 厨房
1. atrium / lobby 2. lounge of emplyee 3. social space 4. open workspace
5. office 6. conference room 7. kitchen
二层 second floor

1 中庭/大厅　2 转接站　3 社交空间　4 开放的工作区间　5 办公室　6 会议室　7 培训室　8 厨房　9 室外厨房
1. atrium / lobby 2. docking station 3. social space 4. open workspace
5. office 6. conference room 7. training room 8. kitchen 9. outdoor terrace
一层 first floor

生态性综合建筑

Triarch Studio

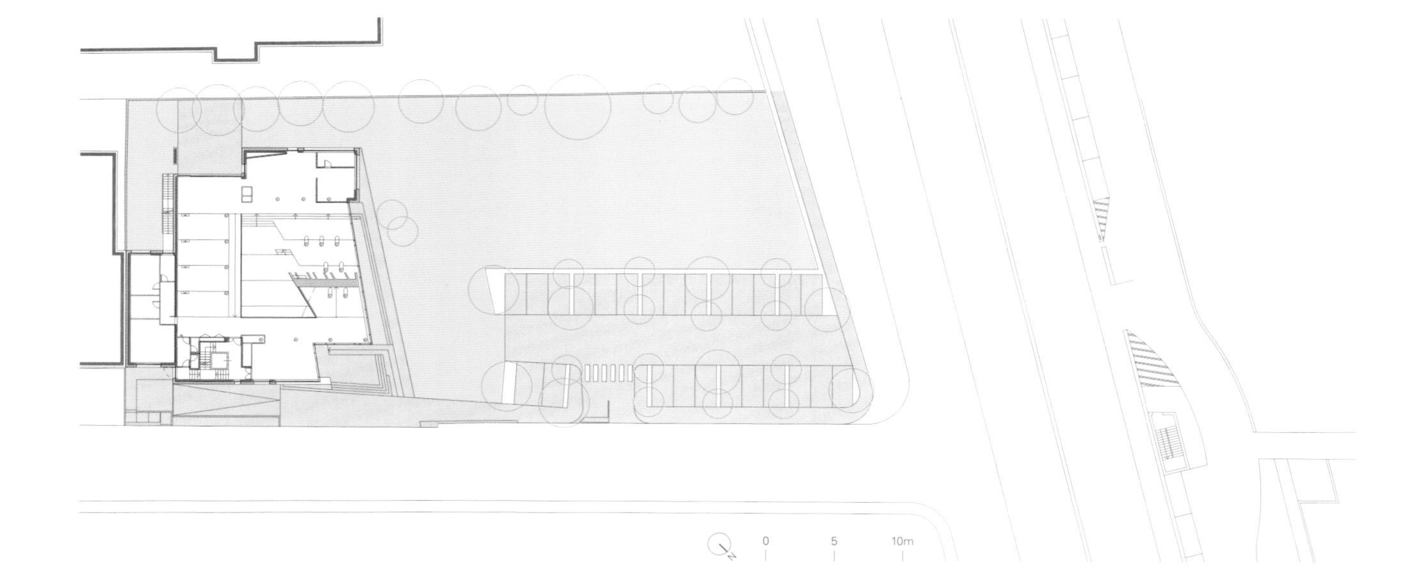

很明显，这座建筑试图将建筑学中的广泛生态原则应用其中。这座综合设施是专门设计用于举办与可持续性和生态性相关的房屋展览、活动以及研讨会，其本身就是如何将绿色原则具体应用于建筑学中的一个典型实例，同时这座设施又有助于当地经济和文化的发展，这也正是这一中心的另外一个主要的商业目标。

该建筑的生态策略包括了许多方面：比如材料的生态足迹、消耗和资源之间的平衡、由公共参与所引起的明显变化而带来的室内舒适度、强大的保温装置（纤维素纤维被放置于外围护结构的280mm空腔中）、大规模自动化以及可再生能量资源的使用。

这个设计试图采用将以下两种主要生态元素相结合的整体策略：自然采光的最优化以及限制直射阳光的处理。由这种整体策略所导致的决策影响了外立面上的实体和空间的使用，也影响了室内空间的功能性布局。玻璃的大小和位置都经过了仔细的计算，并且根据每处建筑区域所接受的直射阳光的总量来决定。除了对整体构成有影响，这种精密的计算也能够使一些区域免受过热的侵扰，同时也能够避免朝北的区域造成过度的热量损失。

实际的建筑结构是多元素的。钢筋混凝土、砖、混凝土柱还有地板都被用于建筑的外环边界中。钢筋混凝土和实心砖幕墙的设计都用于提供足够的体块，以达到一种热惯性，同时，这种设计也将利用传统的当地建筑材料来诠释一种全新的外观。胶合木用于这座建筑的中心区域，在那里有一处很大的上空空间，不仅提供了很好的光线，而且还成为天然的通风烟囱。一条流畅的小道将不同的楼层和夹层连接起来，使人们对这座综合设施有个综览的效果。虽然没有经过划分，但是这座建筑的内部明显地被分为几个不同的部分。倾斜的木柱、斜坡、带透明扶手的楼梯、镶板、露在外面的木梁以及横梁都是这座建筑的其他典型特色。底层的空间是依据其使用功能来划分的，始于接待处，入口大厅则设在底层的西南方向，这里还有一个酒吧/餐厅、一处有着大型门厅和能容纳150人礼堂的会议区，并且侧面展区的中心区域也设于此。展区是运用了“展览会的模块”的理念来设计的，这个理念包含了一个基本的模块，而这个基本模块是可以和其他模块相结合的。这种模型在上层楼面中也被重复利用，使之在中央的上空体量周围形成一个显眼的、完整的螺旋结构，即一处被天窗照亮的连续空间。

空间的利用是很灵活的，所采用的是能够适应各个活动的不同主题的模块选项。实体、围屏以及玻璃之间的互动使这座建筑的立面及其本身变得更具动态性，而体量内的上空空间也使主侧墙这边的入口处和玻璃切口都变得引人注目。

Ecoarea Complex

The building clearly seeks to apply broad-ranging ecological principles in architecture. Designed to house exhibitions, events and seminars related to sustainability and ecology, the complex itself is a clear example of how green principles can concretely be applied to architecture, whilst contributing to local economic and cultural development, since this is another of the center's business goals.

The ecological strategy has numerous parts: the ecological footprint of materials, the balance between consumption and resources, indoor comfort given by the significant variations in public attendance, strong thermal insulation (cellulose fibre was placed in

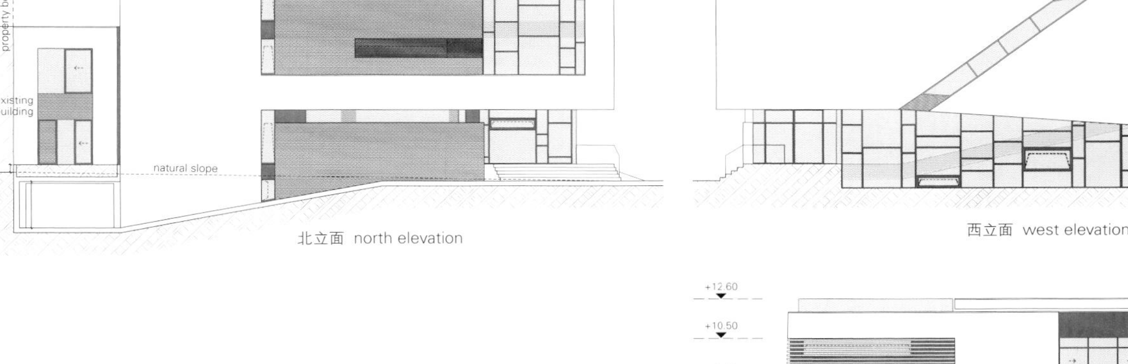

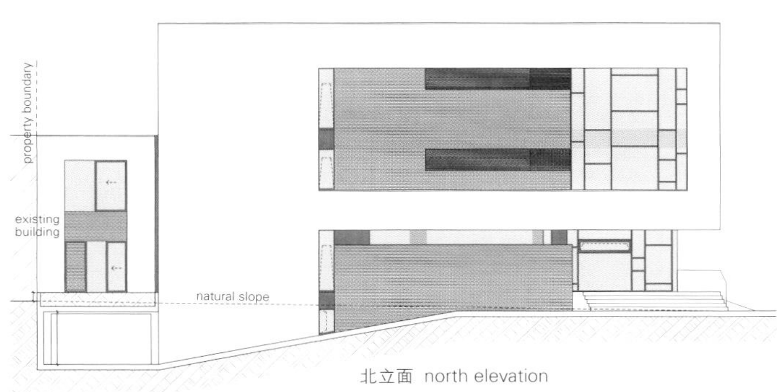

北立面 north elevation

西立面 west elevation

南立面 south elevation

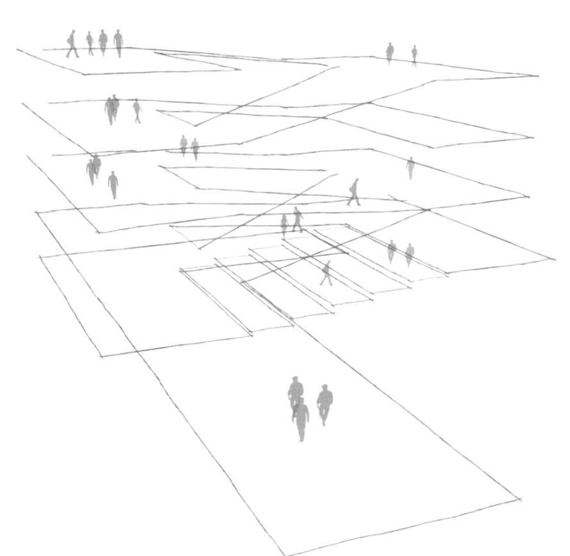

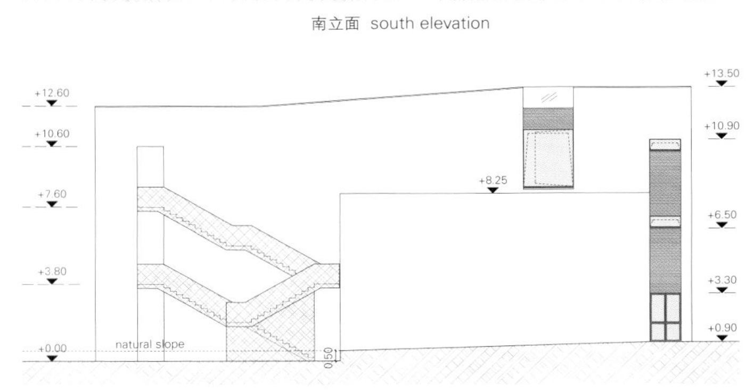

东立面 east elevation

1. 5mm plaster render painted white
   16mm fiber cement panel
   160mm lamellar beams
2. 230mm insulation board made with high density wood fiber
   250mm wall of two bricks thick with lime mortar
   20mm gypsum plaster
3. 130mm polished concrete screed
   5mm rubber acoustic insulation
   280mm reinforced concrete slab
   160mm fiber cement panel
   100mm insulation board made with high density wood fiber
   5mm plaster render painted white
4. roof garden with 180mm earth layer
   50mm gravel for drainage, waterproofing membrane
   140mm insulation board made with high density wood fiber
   280mm reinforced concrete slab
   13mm gypsum board
5. 5mm plaster render painted white
   16mm fiber cement panel
   160mm lamellar beams
   340mm cellulose fibers insulation pumped in the cavity
   250mm wall of two bricks thick with lime mortar
   20mm gypsum plaster

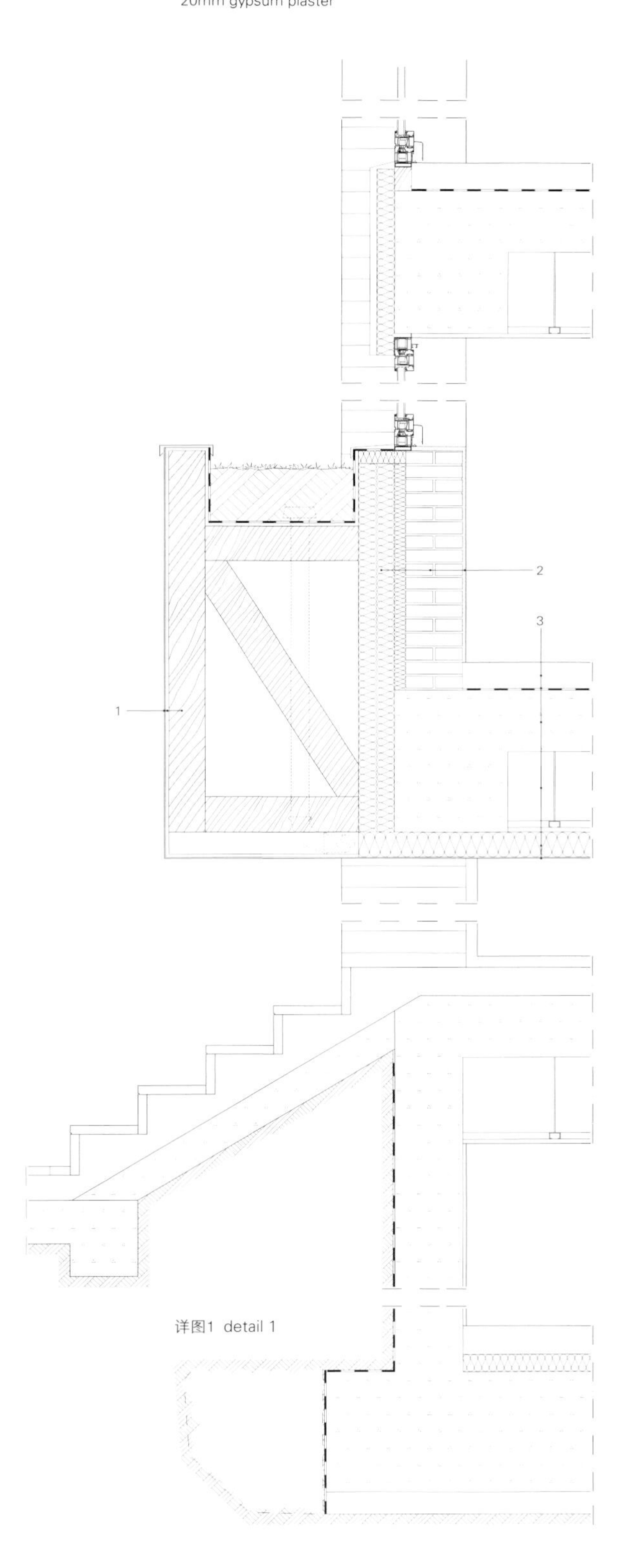

详图1 detail 1

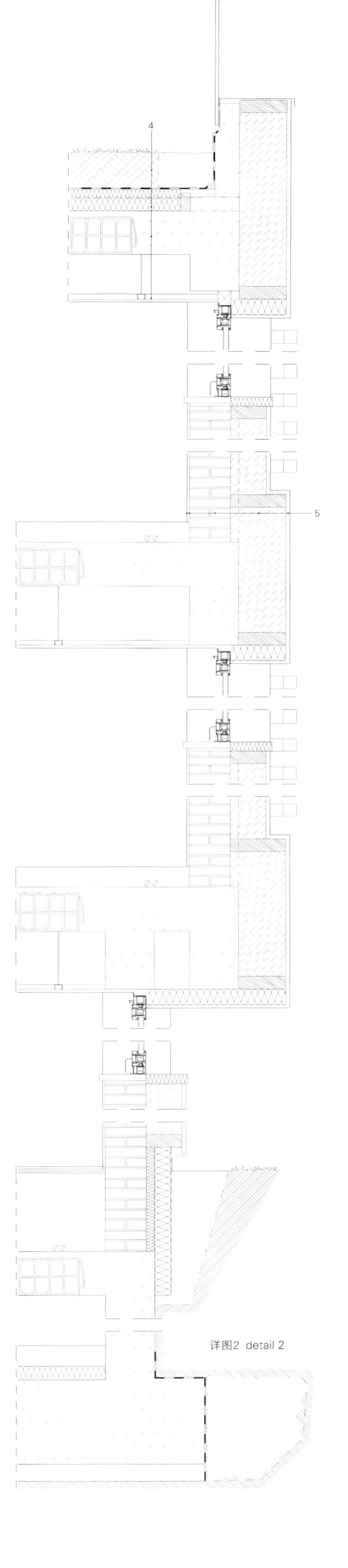

详图2 detail 2

livellozero

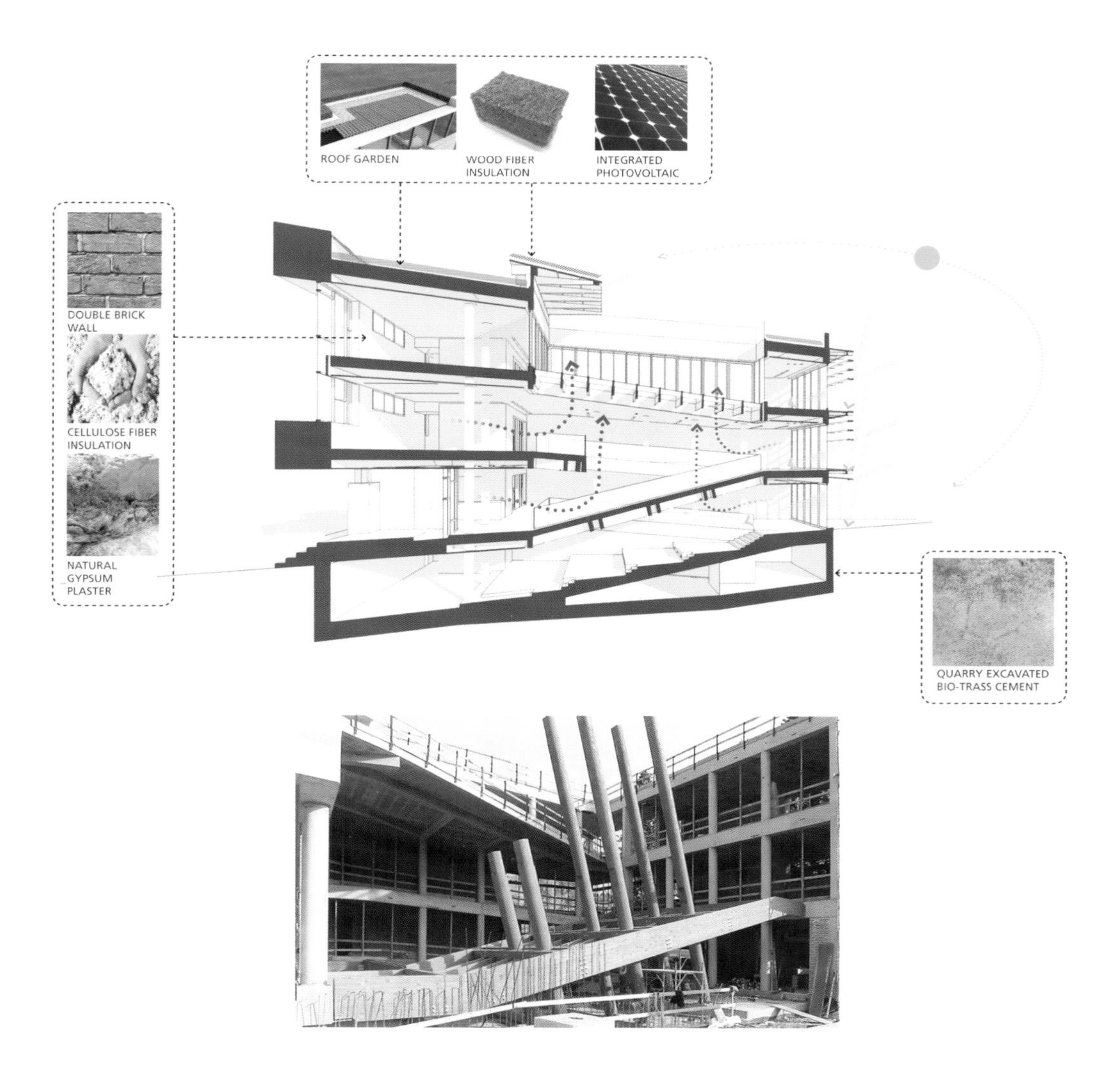

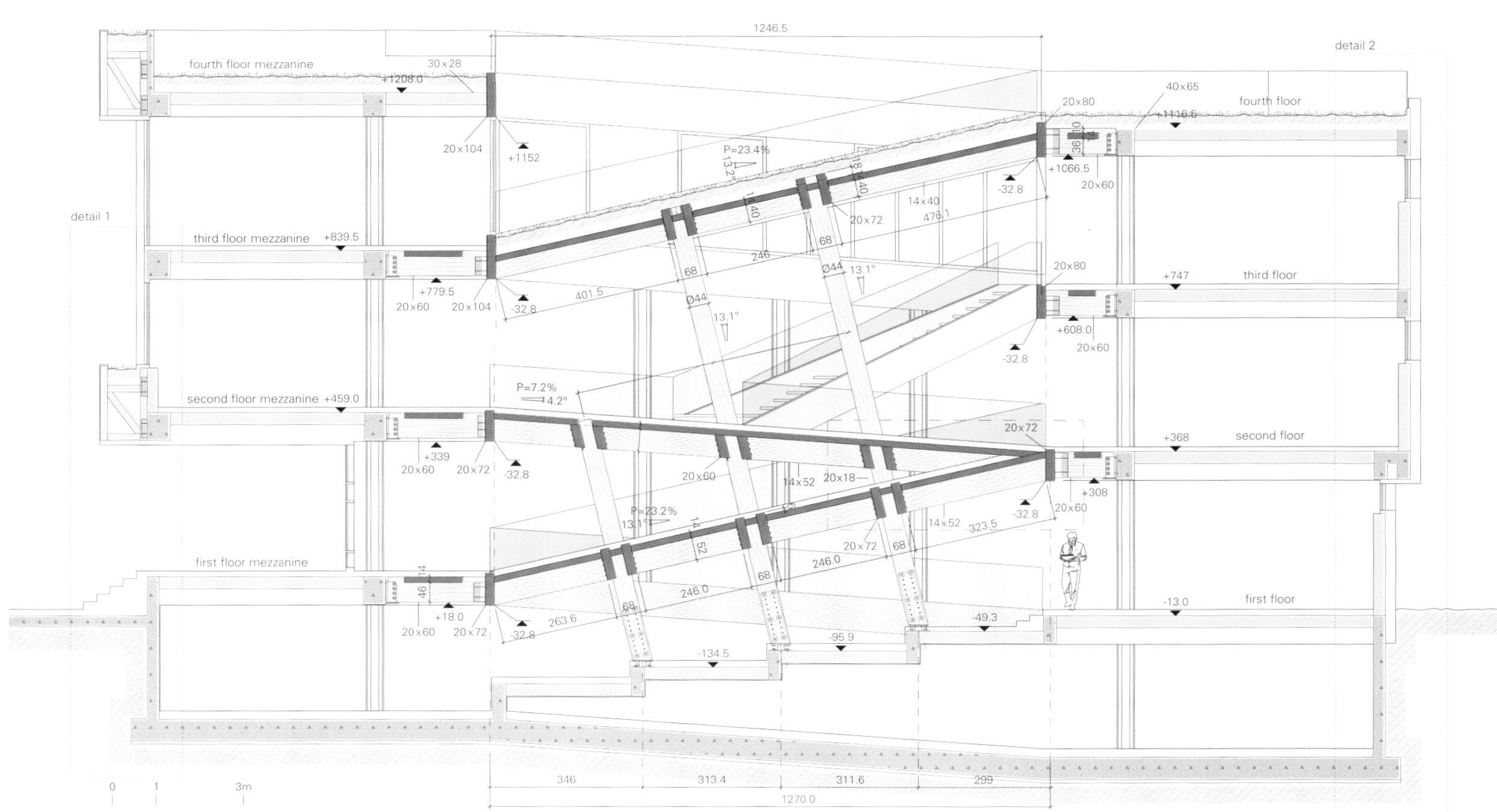

A–A' 剖面图 section A-A'

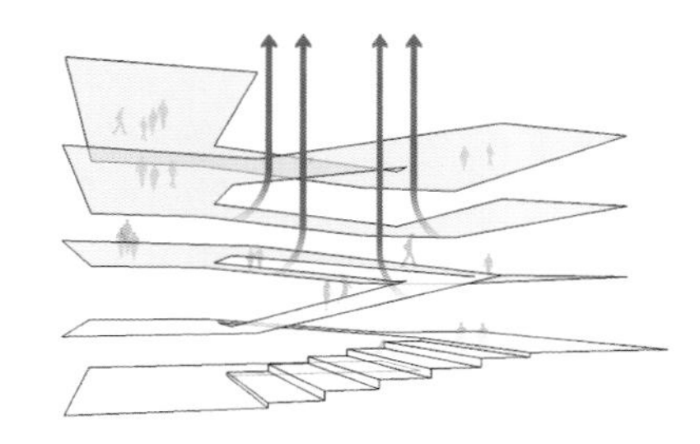

自然冷却通风系统
natural cooling ventilation

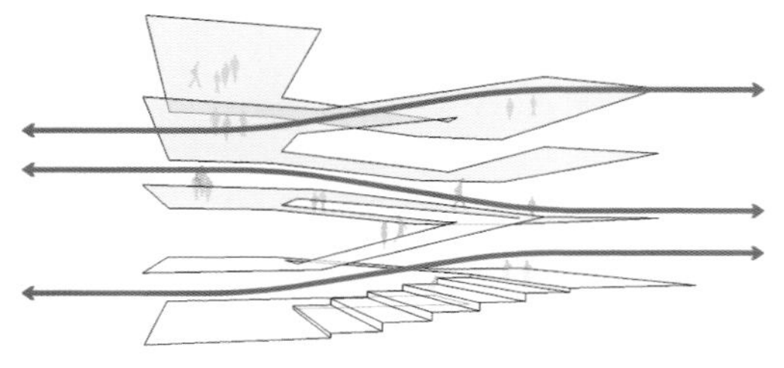

视觉连续性
visual continuity

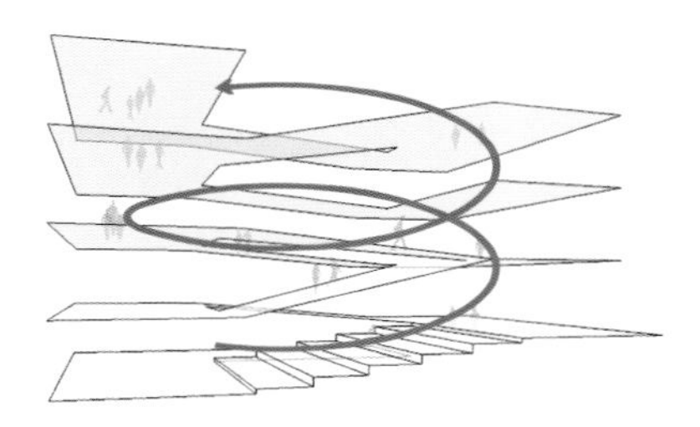

功能连续性
functional continuity

项目名称：Ecoarea
地点：Coriano, Italy
建筑师：Triarch Studio
首席建筑师：Walter Giovagnoli, Alessandro Quadrelli
项目团队：Nerio Tenti, Matteo Maresi, Marco Melucci, Patrizio Giovagnoli, Antonella Fabbri
主要承包商：Carpentedil
轻质技术：ISO3
钢结构：Modelferro Engineering srl
结构工程师：Attilio Marchetti Rossi
热电厂工程：Matteo Pedini
电气工程：Francesco Palmieri
BioTrass混凝土：Holzer
胶合木结构：Habitat Legno
照明：Viabizzuno
窗户细木工：Shuco Artinffissi 2
室内衬里：Fermacell
保温纤维素纤维：Isofloc
甲方：Victoria SAS
用地面积：1,685m²
竣工时间：2012
摄影师：©Gianluca Moretti (courtesy of the architect)

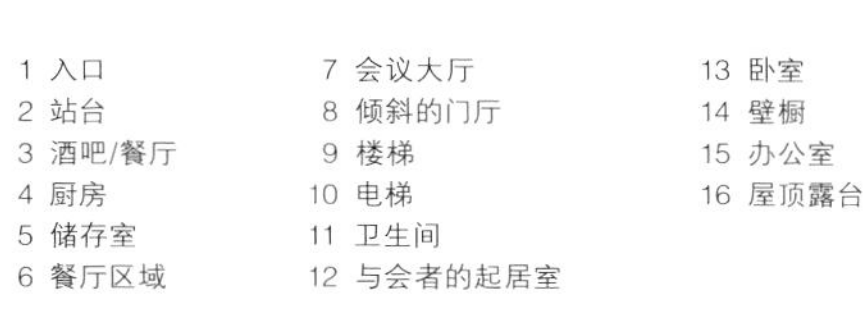

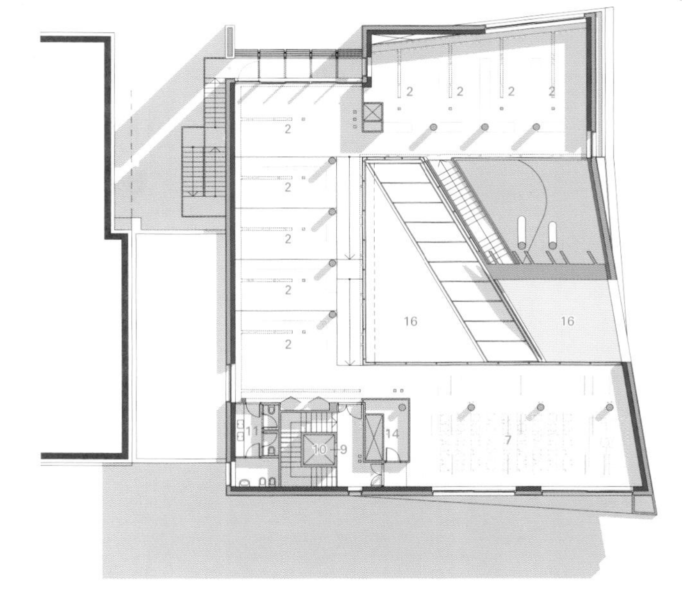
三层 third floor

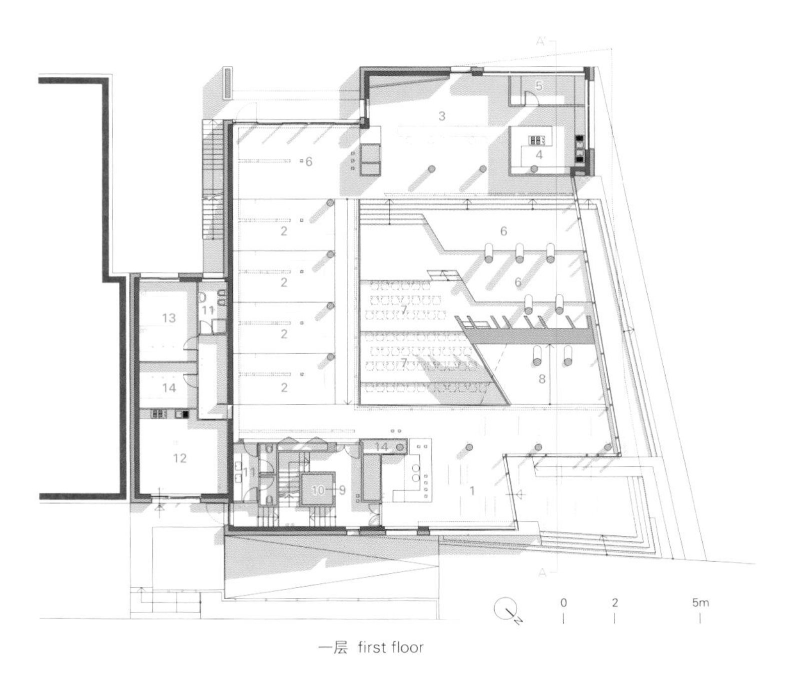
一层 first floor

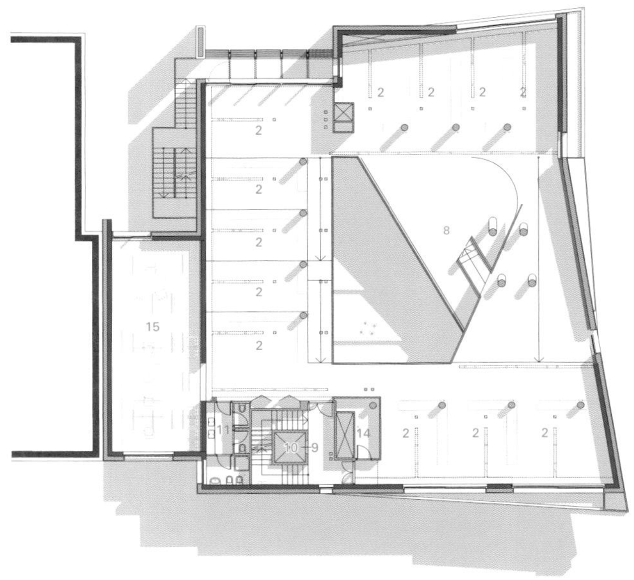
二层 second floor

the 280mm cavity of the envelope), extensive automation and use of renewable energy sources.

The design seeks an integrated approach to merge two key ecological elements: optimising natural lighting and dealing with the constraint of direct sunlight. The resultant decisions affect both the use of solids and voids on the external elevations and the functional layout of internal space. The size and location of the glazing are carefully calculated, depending on the amount of direct sunlight each building section receives. Aside from influencing the overall composition, such calculations also prevent areas from overheating or, on the north-facing side, from excessive heat loss.

The actual structure is mixed. Reinforced concrete is used in the exterior perimeter ring, with brick and concrete pillars and floors. This, along with solid brick curtain walls, is designed to provide sufficient mass to achieve thermal inertia and, at the same time, uses a traditional local building material with a new outlook. Glulam is used in the central section of the building, where a large void dominates, providing both light and acting as a natural ventilation chimney. A seamless path connects the different storeys and mezzanine levels, effectively providing an overview of the complex. The interior is clearly divided into different parts without being fragmented. The inclined wooden pillars, the sloping ramps, the staircase with a transparent parapet, the panelling and the exposed wooden beams and crossbeams are the other defining features. The spaces on the ground floor are divided according to function, starting with a reception area and the entrance hall on the south-west side, a bar/restaurant, the conference area with a large foyer and a 150-seater auditorium, and the core of the lateral exhibition area. The latter has been designed using the concept of "exposition modules" consisting of a basic module that can be combined with several others. This model is repeated on the upper floors, creating a striking unbroken spiral around the central void, a continuous space lit by the skylight.

Space usage is flexible, using modular options that can be adapted to the various themes presented with each event. The elevations and the building itself are made more dynamic through the interplay of solids, screens and glazing, with the void in the volume marking the entrance and the glazed cut in the main side wall.

帕梅拉·科因图书馆

Branch Studio Architects

传统的学校图书馆正朝着越来越数字化的方向发展，而万能的物理书籍也变得越来越稀少了。圣莫妮卡学院图书馆全新的扩建工程由以下两个部分组成：一是翻修了现存的学校图书馆，二是扩建了一处新的阅读区和平台。这个项目通过一系列的中心构思或者是体现建筑嵌入结构的“章节”，来充分表现和展示学校和公共图书馆的传统。而这些嵌入结构作为一个单独的“故事”，来被建筑师集体规划。

1.入口门槛——灵感来源于“秘密花园”的入口。这个入口门槛被规划为一个单独的体量，这也是进入图书馆的一个主要入口。这个门槛巧妙地伸入到现存学校的走廊中，仿佛是通往另一个世界的大门。

2.花园门厅——人们从图书馆的两扇大玻璃拉门出去便来到了庭院，那里有个之前的室内走廊，这个走廊将庭院和图书馆完全隔离开来。图书馆的内部空间和外部庭院共同形成了一处室内或者是室外的阅读区域。

3.“多元服务台”——一个独特的、多功能和多用途的、如“瑞士军刀”般的接待服务台，可用于图书借阅、视听办公轮用、馆藏目录查询，同时也设有休息座位。

4.西班牙式阶梯——现存图书馆的底层楼面和上层楼面之间的1400mm的高度差之前是由两个笨重且狭窄的楼梯连接起来的。现在这个地方已经被重新定义，成为一连串的平台，这些平台可以作为临时的空间，用于讨论、开会、复印资料、阅读、观景，而且人们还可以在这座图书馆的两个不同平面间穿梭。

5.高入树梢的阅读区——一处全新的阅读区穿过一层现存的石墙，延伸至以前废弃的庭院里。这处阅读区是经过精心定位和规划的，以提供朝向附近戴尔滨河岸绿化带以及沼泽地的风光。

这座图书馆还有一系列论坛，供较为小型且更为亲密的学生团体阅读、小组学习，同时还有一对一的教学使用区域。灵活且半透明的员工区鼓励教师以及学生在此活动。发挥分隔作用的窗帘创造出了一处更透明，有时模糊视觉和听觉的屏障空间。同样地，这些窗帘也充当非正式的屏障，以形成灵活的教学区域。当学生在学习单独课程的时候，这些窗帘也许就会被放下来。其他时候，这些窗帘都是拉开的，这样就形成了一处开放的大型学习区域，供午餐时间和放学后使用。

因为胶合板的坚固性和实用性，它被广泛地应用于这座图书馆的建造中。学校的操场已经有大量磨损的地方，有了胶合板这种材料，这些划痕和刮伤反而会增强这种胶合板的光泽度。垂直的天然木板条和染黑的胶合板覆层增强了内外空间之间的视觉连接。

这座建筑是完全隐于学校外面的。坐落于天然的树冠内，这座全新的建筑只能从旁边的人行道窥见一斑，几乎和周围的环境融为一体。

Pamela Coyne Library

The traditional school library is becoming more digitalized and the all mighty physical book is becoming more and more scarce. The St Monica's College Library's new extension consists of two parts: 1. A renovation to the existing school library and 2. A new reading lounge & deck extension. The project celebrates and elaborates on the traditions of the school & civic library through a series of key ideas, or "chapters", that were translated into architectural interventions. These architectural interventions were collectively composed and narrated as a single "story".

1. Entry Threshold – Inspired by the entry to the "Secret Garden", the Entry Threshold is conceived as a singular volume and is the main entry to the library. The threshold protrudes slightly into the existing school's corridor like a portal into another world.

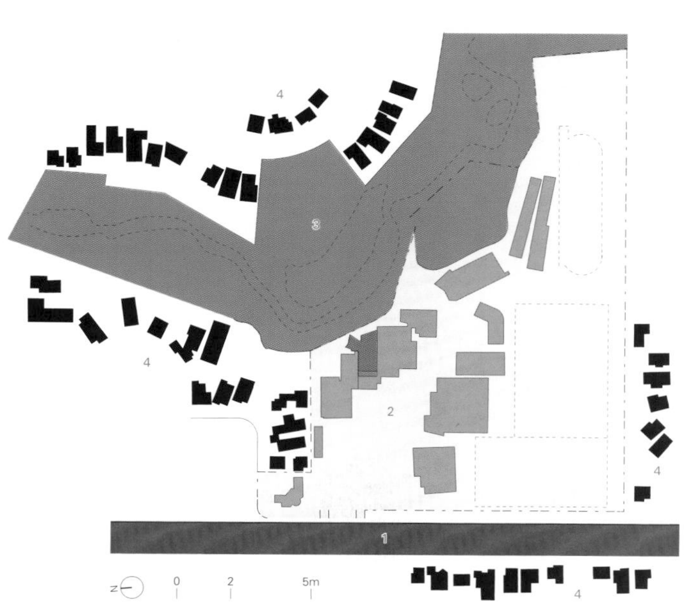

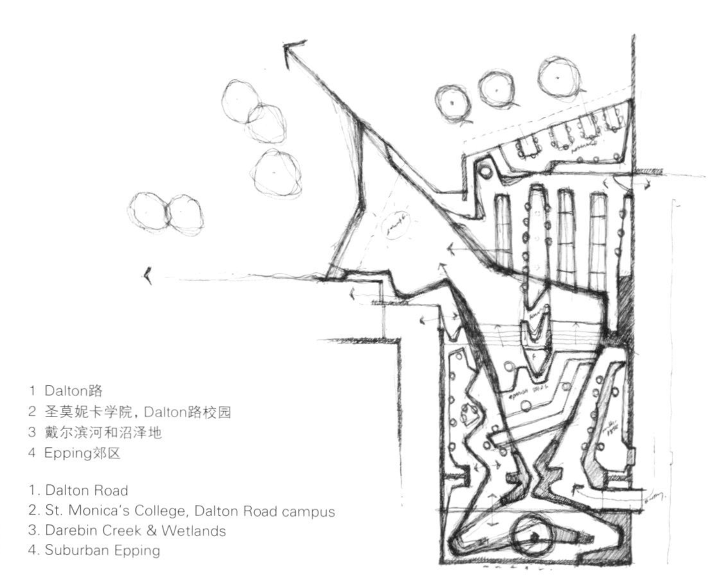

1 Dalton路
2 圣莫妮卡学院，Dalton路校园
3 戴尔滨河和沼泽地
4 Epping郊区

1. Dalton Road
2. St. Monica's College, Dalton Road campus
3. Darebin Creek & Wetlands
4. Suburban Epping

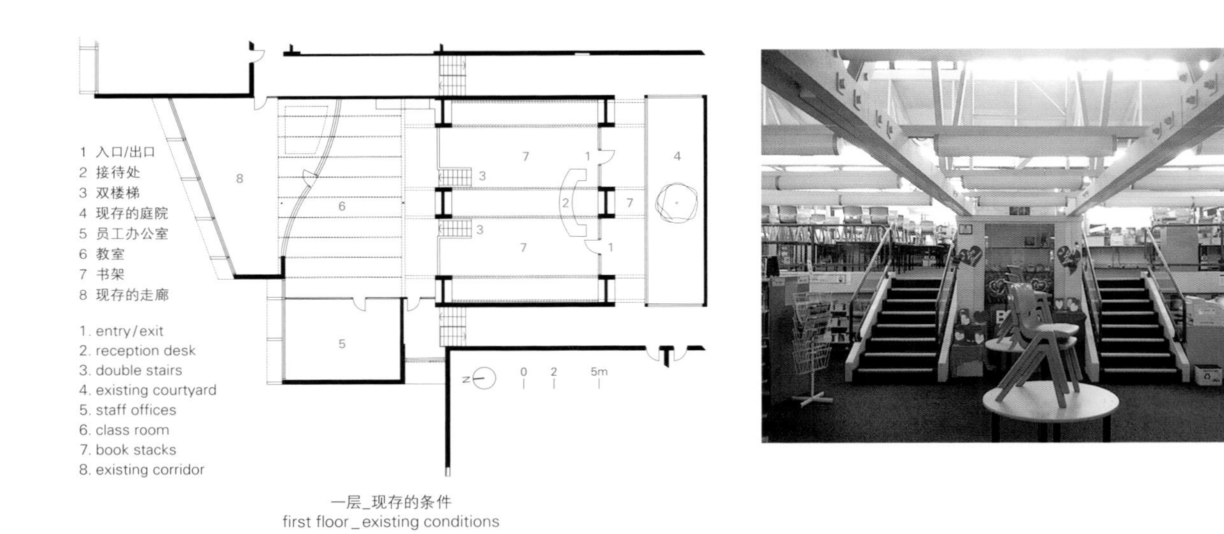

一层_现存的条件
first floor_existing conditions

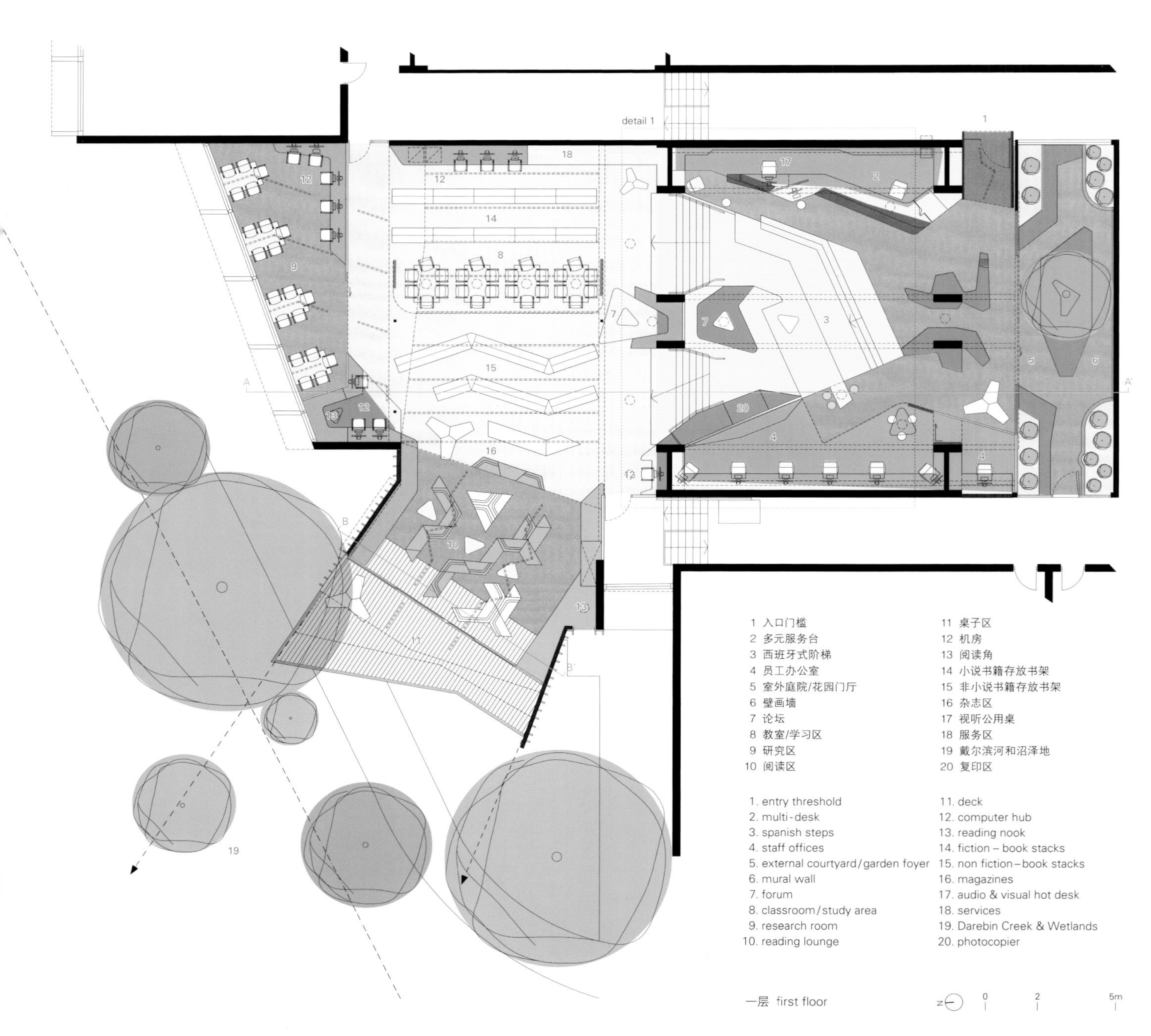

一层 first floor

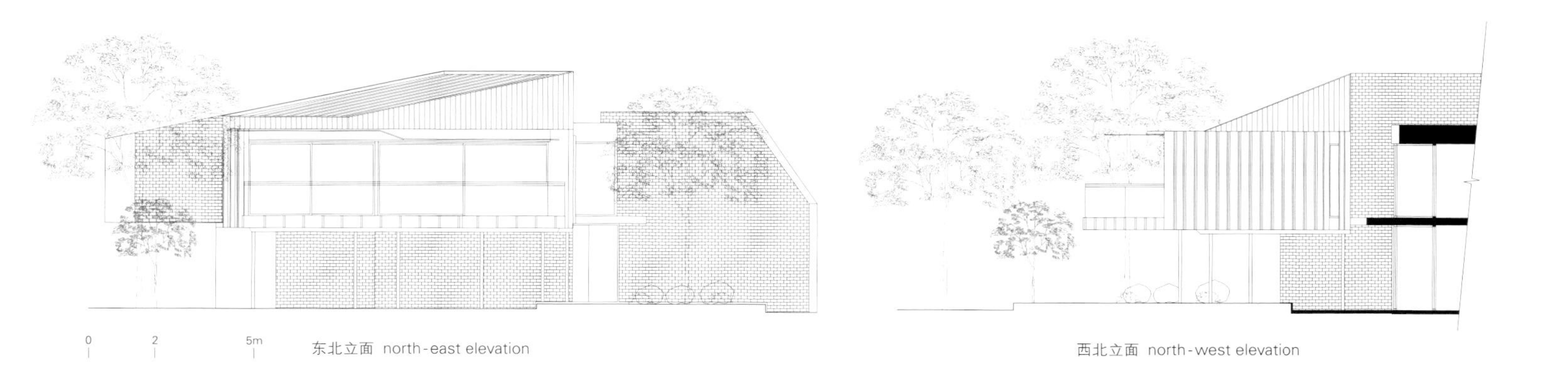

东北立面 north-east elevation

西北立面 north-west elevation

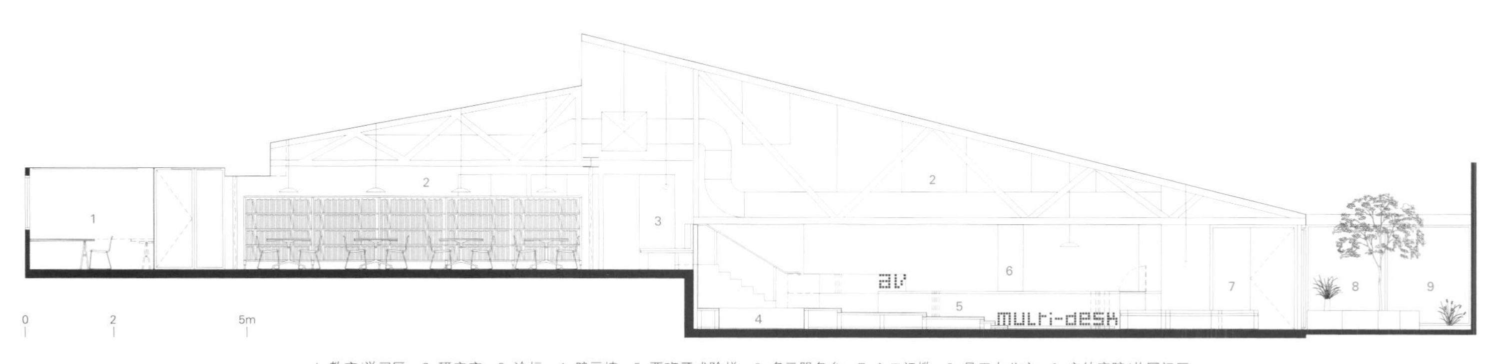

1 教室/学习区　2 研究室　3 论坛　4 壁画墙　5 西班牙式阶梯　6 多元服务台　7 入口门槛　8 员工办公室　9 室外庭院/花园门厅

1. classroom/study area 2. research room 3. forum 4. mural wall 5. Spanish steps 6. multi-desk 7. entry threshold 8. staff offices 9. external courtyard/garden foyer

A-A' 剖面图 section A-A'

详图1_西班牙式阶梯
detail 1_spanish steps

2. Garden Foyer – Two large glass sliding doors open up the library to the courtyard where an existing internal corridor is used to segregate the courtyard from the library completely. The internal library spaces are now engaged with the external courtyard, creating an indoor/outdoor reading area.
3. The "Mutli-desk" – It's a singular multi-purpose, multi-use, "Swiss army" reception desk, catering for borrowing, audio-visual hot-desk, library catalogue and a seat.
4. The Spanish Steps – A existing 1400mm change in levels between the lower & upper floors of the library was previously connected by two awkward, narrow stairs. This has been redefined as a series of platforms that promote impromptu spaces for discussion, meeting, photocopying, reading, viewing and traversing between the two levels of the library.
5. Tree-top reading lounge – A new reading lounge punches through an existing brick wall on level 01 and extends out over a previously unused courtyard. The reading lounge is specifically orientated & configured to offer views towards the nearby Darebin Creek's green belt and wetlands.
The library contains a series of Forum spaces for smaller, more intimate student reading, study groups and area for one-on-one teaching. Flexible & translucent staff areas encourage teacher's and student's engagement. Through the use of a curtain divider, a more transparent & sometimes blurred visual & spatial barrier is created. Similarly, curtains are used as informal screening devises to create flexible teaching & study areas. When a private class is required curtains may be pulled shut. At other time the curtains can be pulled open for the area to be used as a large open study area during lunchtime and after school.
Plywood was used generously throughout for its durability and practicality. The schoolyard is a place where wear and tear is common and plywood is a material where scratches and scuffs could add to the patina of the material. Vertical natural timber battens, in collaboration with black stained plywood cladding, promote a visual connection with internal and external spaces.
The extension is almost completely hidden from outside of school. Situated amongst the tree canopies with only a small glimpse to be seen from a nearby walking track, the new building's fabric merges with its surroundings.

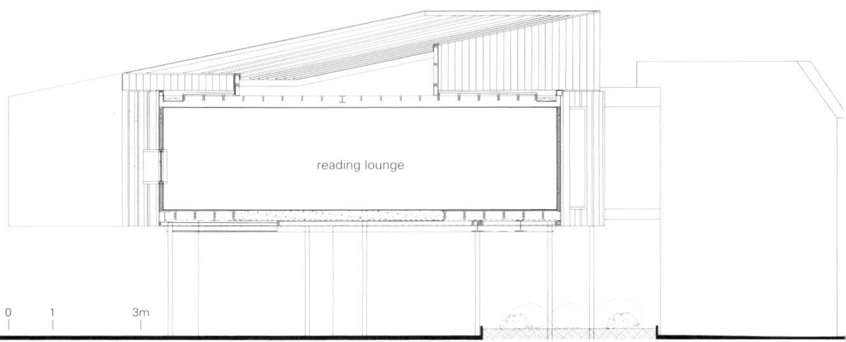

B–B' 剖面图 section B-B'

项目名称：Pamela Coyne Library
地点：St Monica's College, 400 Dalton Road, Epping, Victoria, Australia
建筑师：Branch Studio Architects
项目团队：Brad Wray
结构&土木工程师：Perrett Simpson Stantin
机械&电力工程师：JBA Consulting Engineers
景观建筑师：Branch Studio Architects, St Monica's College
土地测量师：MJ Reddie
低值易耗的家具设计：Branch Studio Architects, Furniture Concepts
甲方：St Monica's College
新扩建的总建筑面积：38m²
有效楼层面积：501m²
设计时间：2013 / 施工时间：2013
竣工时间：2014
摄影师：©Nils Koenning (courtesy of the architect)

绿色能源实验室

Archea Associati

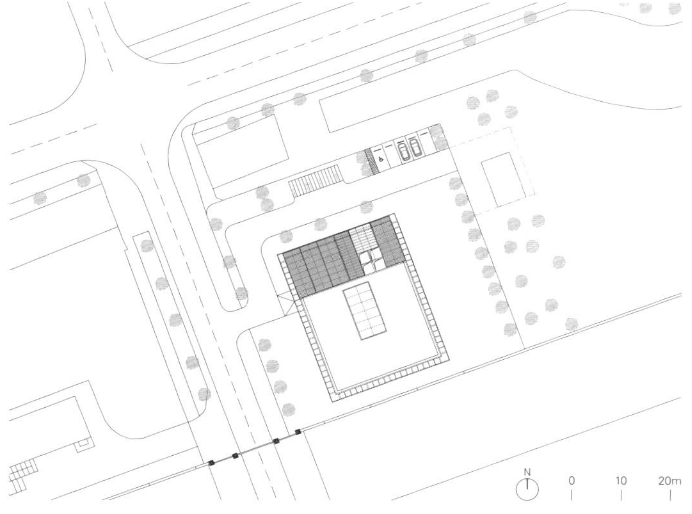

在2007年11月，上海交通大学和意大利环境和海陆保护部门签署了一份合作协议，一起建成了GEL，即"绿色能源实验室"的简称。这座建筑由阿克雅建筑师事务所设计，法维罗&米兰工程公司规划其结构，并在上海交通大学的闵行校区内建造。

这个实验室于2012年4月竣工，在同年的5月19日开始正式开放，环境部长科拉多·克利尼也出席了这个开幕式。GEL用作施工过程和住宅使用中应用的低碳排放技术的研究中心和实验室，被构思为一个紧凑的体量，围绕在一个庭院的周围，其上方被一个大型天窗覆盖，天窗可以根据季节的变化打开或者关闭，这个方案是根据其布局和能源最优化的功能性特点而做出的。

这个被过道所环绕的空间被构想为一个能将能量消耗最优化的体量：在冬天有阳光的时候，这处空间就可以作为一个储热器，而在夏天的时候，它又可以作为一个烟囱，将室内的热气排放出去。这座建筑一共有三层，地上总表面面积为1500m²，最高高度为20m。实验室、会议

西立面 west elevation

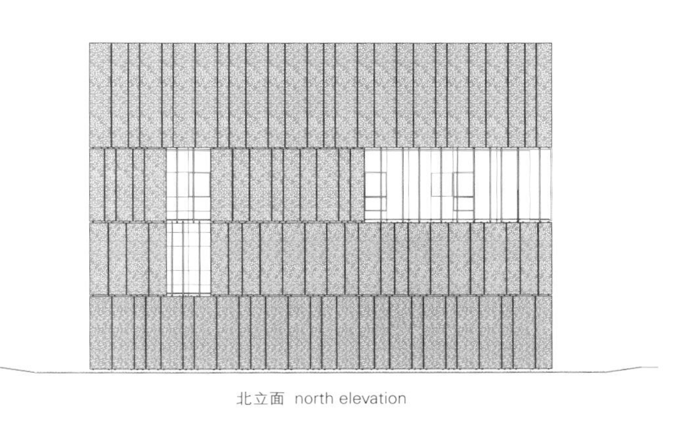

北立面 north elevation

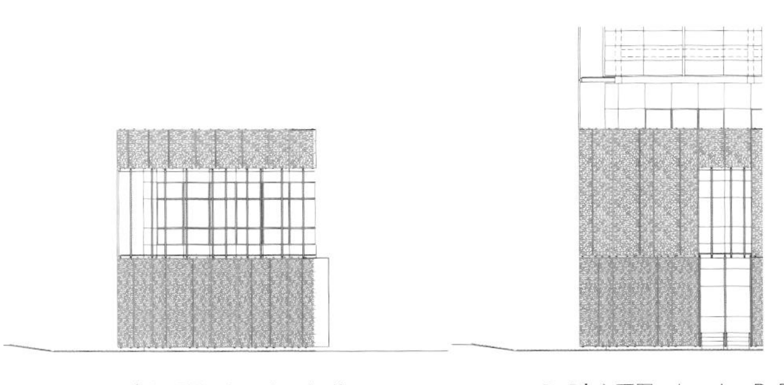

A-A' 立面图 elevation A-A'

B-B' 立面图 elevation B-B'

室、控制室、学生教室以及展览区都在一层和二层；每处室内空间的两侧都设有窗户，透过窗户可以看到外面的空间以及内部的庭院。三层设有两个样板房，模拟了一个两居室和一个三居室的公寓，覆盖了带有光生伏打板的斜屋顶，斜屋顶作为平台，可用于测量居住空间的类型是否适合，以对节能系统和建筑进行测验。这座建筑的朝向、矩形的外观、立面以及内部玻璃围合的庭院的设计都能最大限度地利用自然通风并且控制其曝光程度，目的是以最低的能量消耗使室内达到一个理想的温度。这座建筑的立面是其外部体量的一个典型特色，且含有双表皮：玻璃单元的内层起到了防水和绝缘的作用，而外层则是一个瓷质百叶，相当于一个遮光板，起到了遮挡和调节内部工作间的照明的作用。暖通空调系统是基于一个中心系统（CHPC/WHP）和其他专用的小型系统来设计的，这些小型系统可以根据不同实验室的试验和研究结果来互相通用。

Green Energy Laboratory

The cooperation agreement signed in November 2007 between the Shanghai Jiao Tong University and the Italian Ministry of the Environment and Protection of the Territory and the Sea has resulted in the construction of the GEL, acronym of the "Green Energy Laboratory". The building, designed by Studio Archea Associati in collaboration with the engineering firm Favero & Milan which has planned the structures, has been built in the Minhang campus of the Jiao Tong University.

It was completed in April 2012 and opened on the following 19 May 2012 in the presence of the Minister of the Environment, Corrado Clini. Created as research center and laboratory for the analysis and diffusion of low carbon emission technologies in the construction and housing sector, the GEL is conceived as a compact body surrounding a central court, covered by a large skylight that

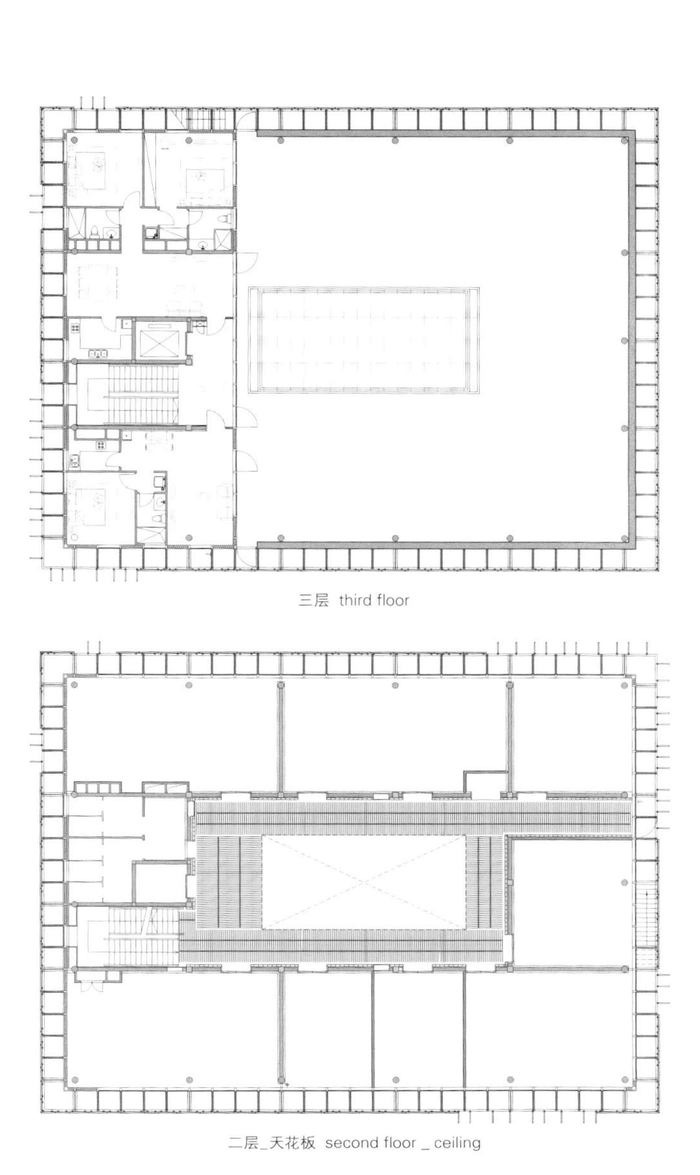

三层 third floor

二层_天花板 second floor _ ceiling

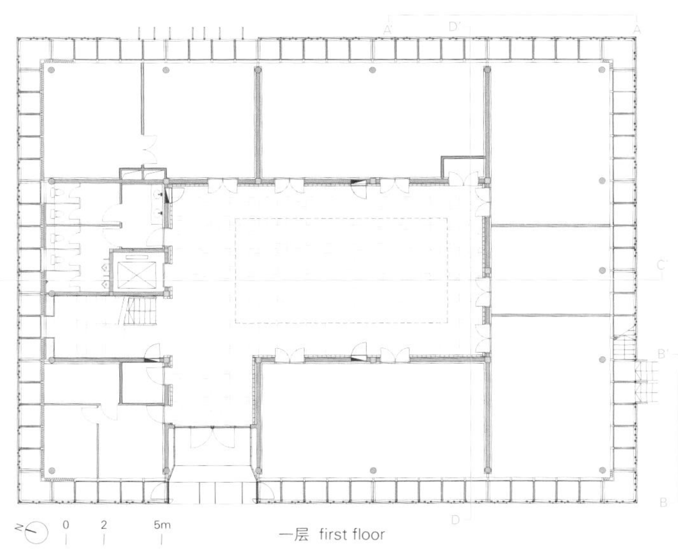

一层 first floor

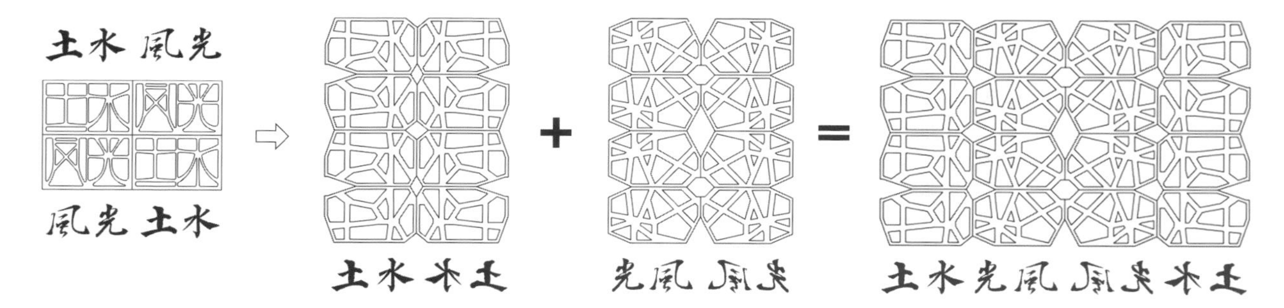
土水 風光
風光 土水
土水
光風
土水 光風

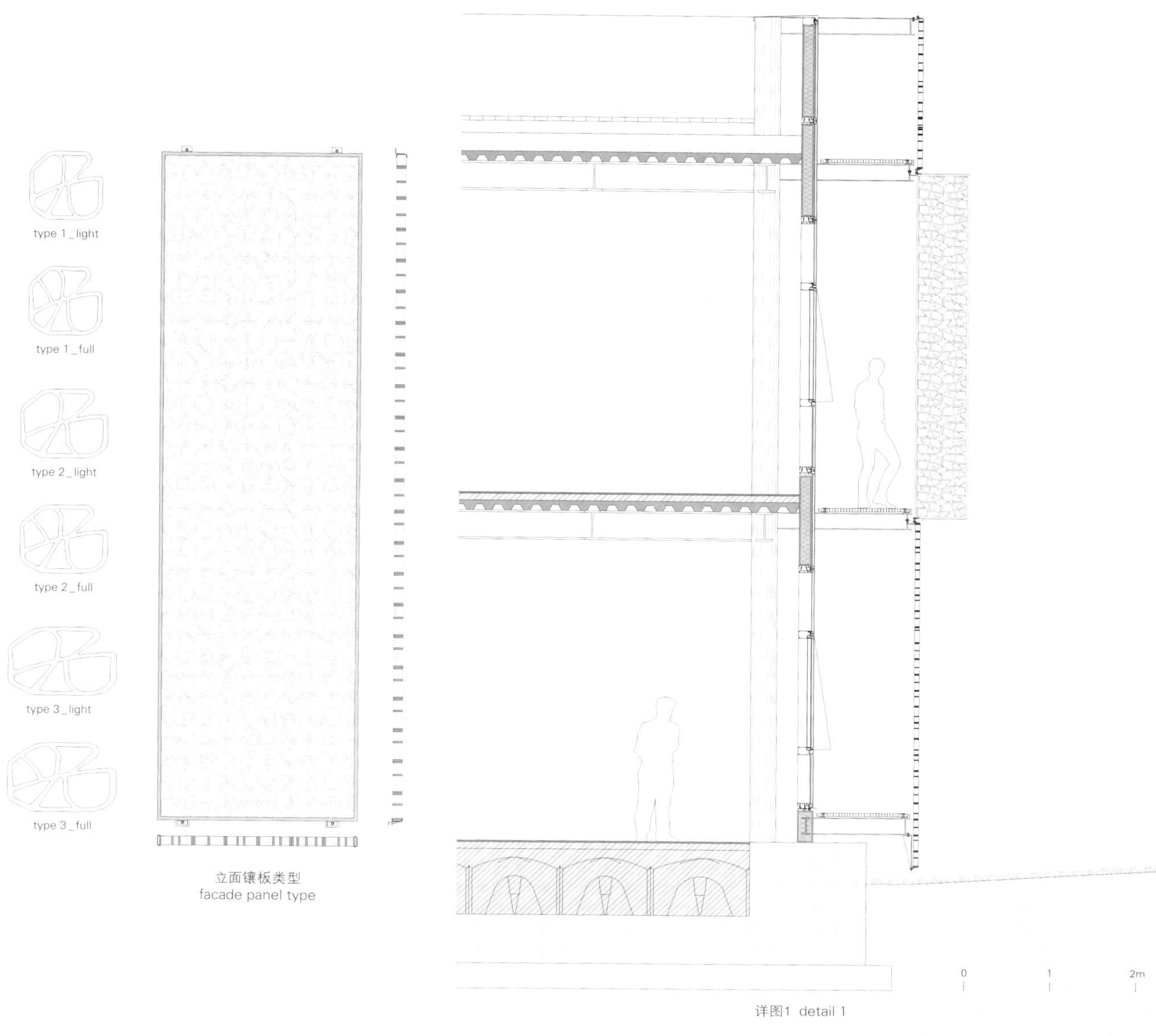

立面镶板类型
facade panel type

详图1 detail 1

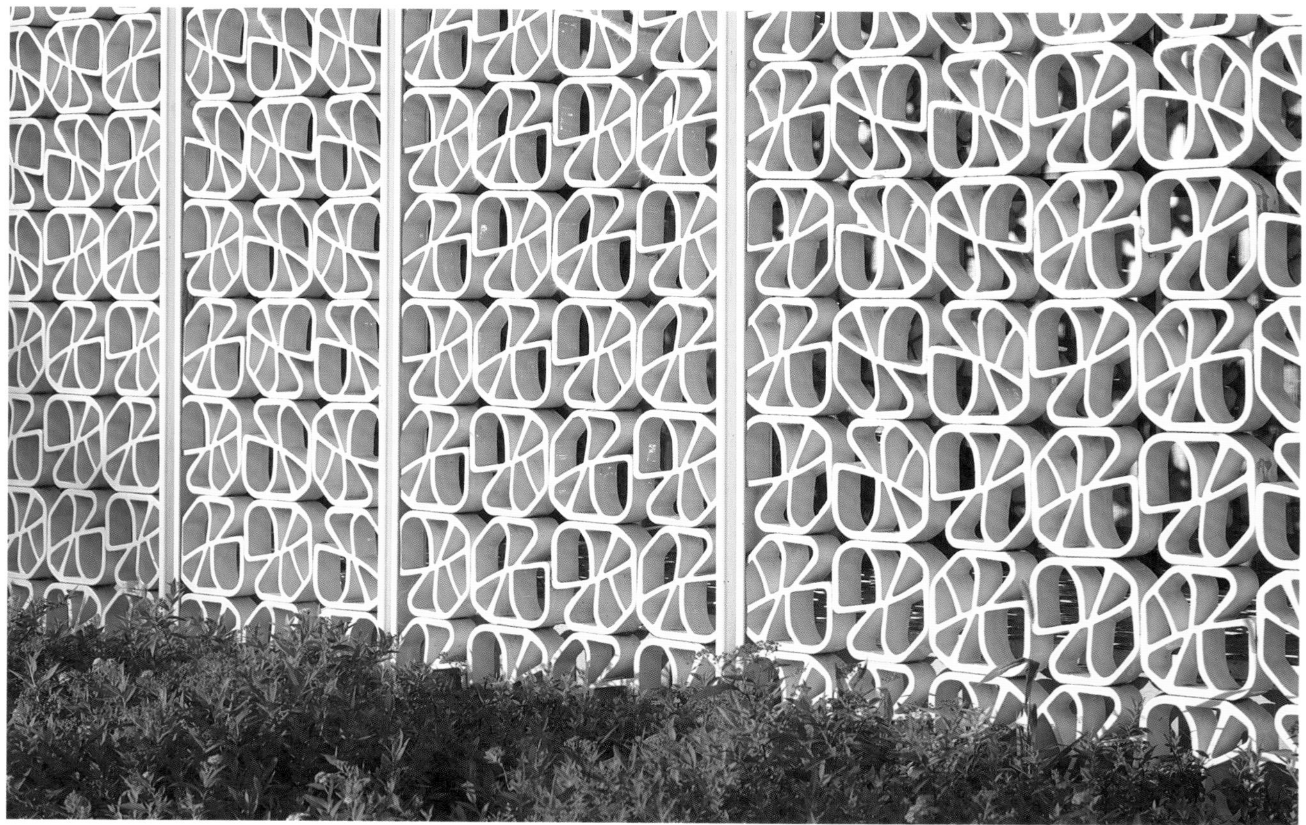

can be opened or closed depending on the season, a solution chosen due to its functional characteristics in terms of distribution and energetic optimization.

The space, surrounded by access balconies, is configured as a void that optimizes energy consumption; on sunny winter days it functions as an accumulator of heat, and in summer it acts as a chimney, aspirating the hot air produced in the interior. The building has three floors with a total surface area of 1500 square meters above ground, and a maximum height of 20m. The first two floors host laboratories, meeting rooms, a control room, classrooms for the students and an exhibition space; every interior has windows on two sides, to the exterior and the inner court. The third floor hosts two sample apartments, the simulation of a two-room flat and a three-room flat covered by a pitched roof with photovoltaic panels, realized as the platform for tests on residential types of spaces, to experiment with energy-efficient systems and buildings. The orientation of the building and its rectangular shape, along with the facade and the glazed interior court, are conceived to maximize the natural ventilation and to control exposure to the sun, in order to obtain an ideal interior climate with a minimum expenditure of energy. The facade, the distinctive feature of the exterior volume, consists of a double skin: an internal layer in glazed cells that provides waterproofing and insulation and an external one consisting of earthenware shutters that serve as sunscreens, to shade and regulate the illumination in the working spaces inside. The HVAC system has been designed on the basis of a main system(CHPC/WHP) combined with other dedicated ones of smaller dimensions that are interchangeable according to the tests and research work done in the different laboratories.

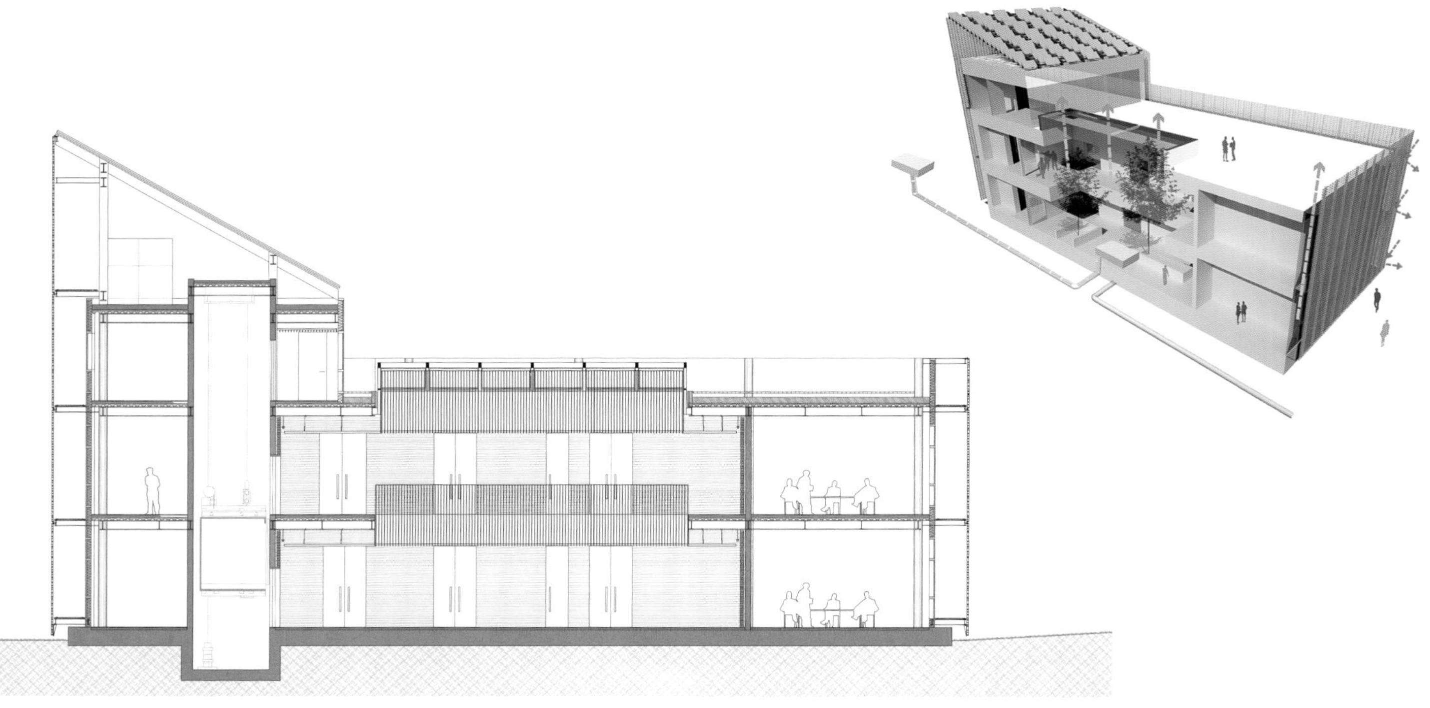

C-C' 剖面图 section C-C'

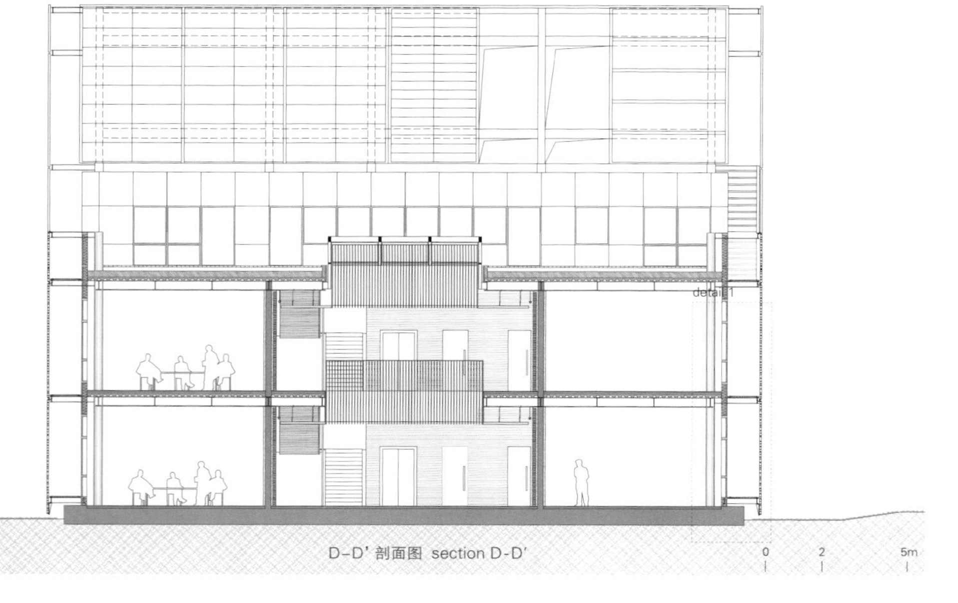

D-D' 剖面图 section D-D'

项目名称：GEL-Green Energy Laboratory
地点：Minhang Campus, Jiao Tong University, Shanghai, China
建筑师：Archea Associati
项目团队：Laura Andreini, Marco Casamonti, Silvia Fabi, Giovanni Polazzi
项目经理：Enrico Ancilli
场地协助员：Andrea Antonucci, Wang Xinfang
结构工程师：Favero&Milan ingegneria S.r.l.
系统：TIFS Ingegneria
认证：LEED Golden Category
甲方：Jiao Tong University
功能：research center
场地表面积：1,500m²
建筑表面积：4,850m²
体积：27,000m³
设计时间：2008
竣工时间：2012
摄影师：©Charlie Xia (courtesy of the architect)

雷普索尔园区

Rafael de La-Hoz Arquitectos

这座建筑地处门德斯·阿尔瓦罗，项目的基本理念是为了在马德里的市中心建造一处商业园区。建筑所占用的地界面向场地中心的一座景观花园开放。

这座全新的总部大厦将由四栋建筑组成，中间建有一个庭院，庭院将是整个设计的一个亮点。阶梯式迂回的体量构成既坚固耐用，又十分显眼。其主要部分展现了这些特有的楼层是如何通过钢梁来建造的。这些钢梁有一个非常薄的，仅有20cm的剖面，使窗框得以放入其中。整座建筑的有效楼层面积约为130 000m²，总用地面积为32 000m²，包含了四栋建筑，而且内部将会修建公共空间和大众观景区域。这座全新的建筑将容纳4000人，并且拥有一个能容纳2000台车辆的两层地下停车场。

开放空间的中心是一个占地10 000m²的中央庭院，庭院中郁郁葱葱的成年天然松树能进行遮阴。这个庭院不仅提供了与人会面、放松身心，以及在25m宽的水池里尽情享受凉爽的机会，还能使4000多名员工观赏到升起的种植池里的装饰性植物。在这座建筑的上层楼面，攀岩植物遮蔽的阳台形成了亲密的室外空间氛围，并且在西班牙炎热的夏天，可以改善室内的微气候。所有元素的设计都与建筑的结构相对应：大型铺路板、种植池、水池以及树木。植被种类的选择是与气候，光反射材料以及蓄水池的水管理相匹配的，而且良好的照明有助于满足可持续性的LEED高标准。

可持续性场地

这座建筑的选址使之前的一个工业区和现有的基础设施得以进一步发展。这里有着广泛的公共服务网络（公车、地铁、当地铁路、高铁等等），而且强大的警力也使这里的人们愿意骑自行车，电动车，以及使用容量大、低排放、节能型的交通工具。建筑公司也通过一系列的行动来与附近社区形成良好的关系，比如用非食用水来淋湿建筑场地，目的是避免尘土遮天现象的发生；在离开前清理卡车留下的轮胎印，使整个区域形成人行道专区，以及在施工期间给附近的居民提供热线服务电话。

节水性

设计很少需要浇灌的花园，选择能够适应马德里气候的植被，使用节水装备（浴室、水龙头、喷头等）以及使用雨水来浇灌（这些雨水是储存在一个容积为250 000L的地下水槽中），这些节能设计都使水资源消耗最优化。

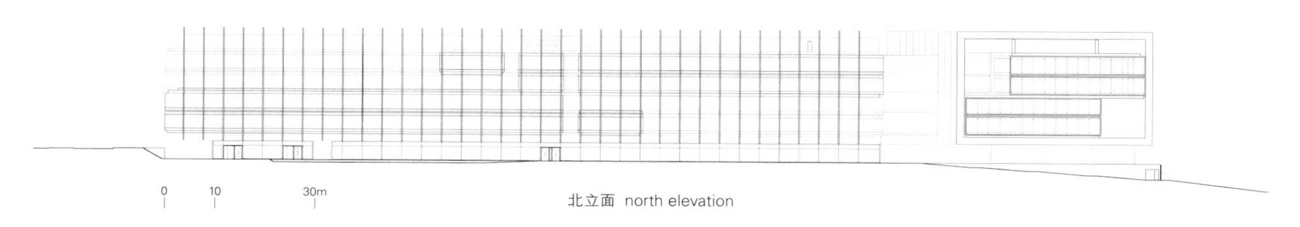

北立面 north elevation

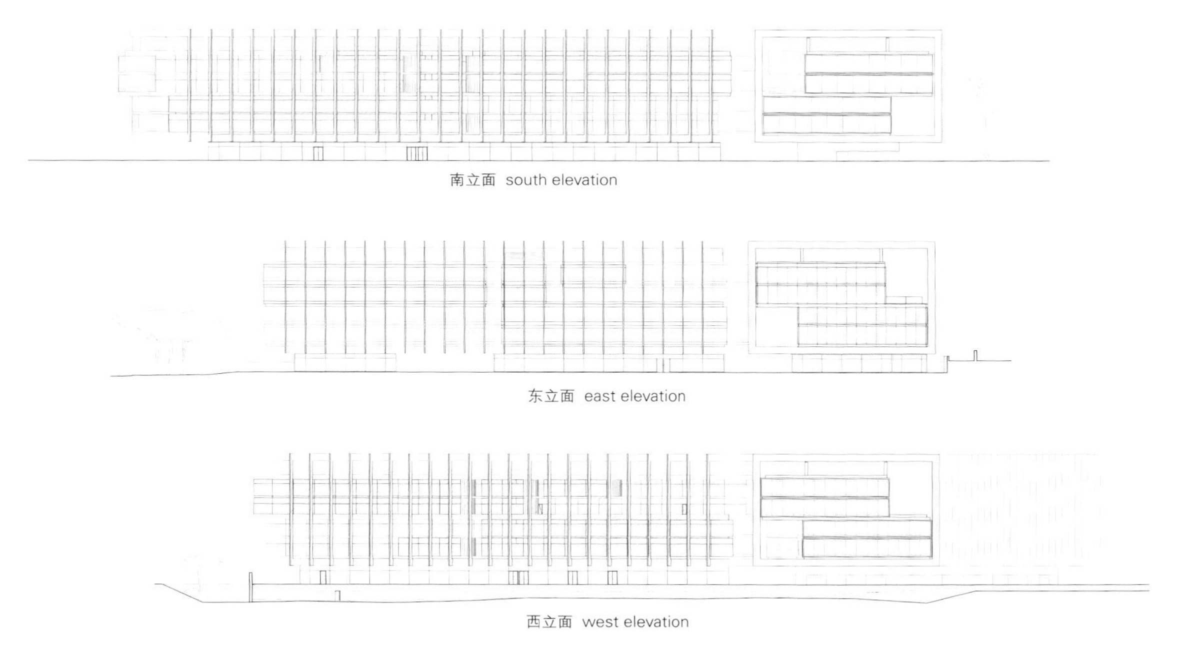

南立面 south elevation

东立面 east elevation

西立面 west elevation

能源和空气

设计师找到了交替能源发电装置，如用于控制气候和产生热水的光伏太阳电池板或者是气体加热泵。

材料和资源

从目前的设计阶段来看，项目选用的是高度再生型材料，这样可以减少原材料在提取和加工过程中的消耗，并且支持当地资源的使用，鼓励精心的森林经营。

室内环境质量

除了将碳排放最小化以外，这座建筑还通过控制照明系统、最大限度地利用室内自然光以及为住户提供一个欣赏花园美景的绝好视角，来使住户们意识到他们拥有一个良好的居住环境。通过使用能够监测可呼吸的空气质量的最先进的气候调节系统，整座建筑确保了一个舒适的温度。

创新设计

交通管理计划的实施为人们提供了出行的多重选择和替代选择，这样便大量地减少了私家车的使用。

Campus Repsol

Located in Méndez Álvaro, the basic idea of the project is to build a business campus in the heart of the city of Madrid. The buildings occupy the perimeter which opens up space for a landscaped garden in the center of the plot.

The new headquarters will be composed of four buildings allocated around an interior courtyard that will be one of the key points of the project. The composition of stepping forth and back volumes creates a strong and expressive volumetric play. The main section shows how the typical floors are framed by steel beams. The steel beams have a very thin section of 20cm which allows the boxes to be appreciated. The entire construction contains a gross floor area of around 130,000m2 which will be placed on a plot area of 32,000m2 and will have four buildings, and the interior will have shared spaces and common landscaped areas. The new building will have the capacity to hold 4,000 people and a two-story underground car park for 2,000 vehicles.

Core of the open spaces is the 10.000m2 central courtyard shaded by mature native pine trees. It offers various opportunities to meet, relax or savor the coolness of the 25m wide water pool or

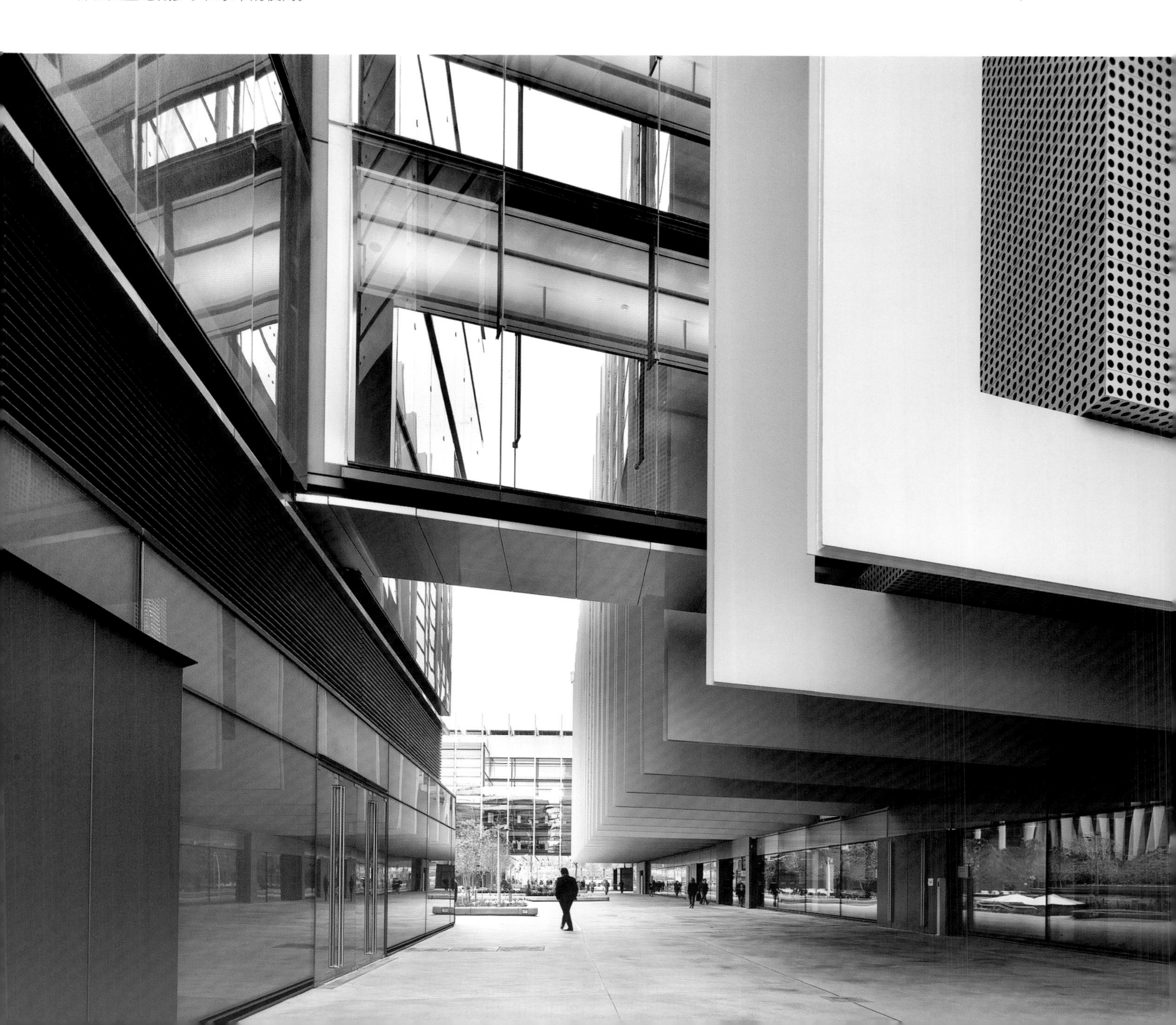

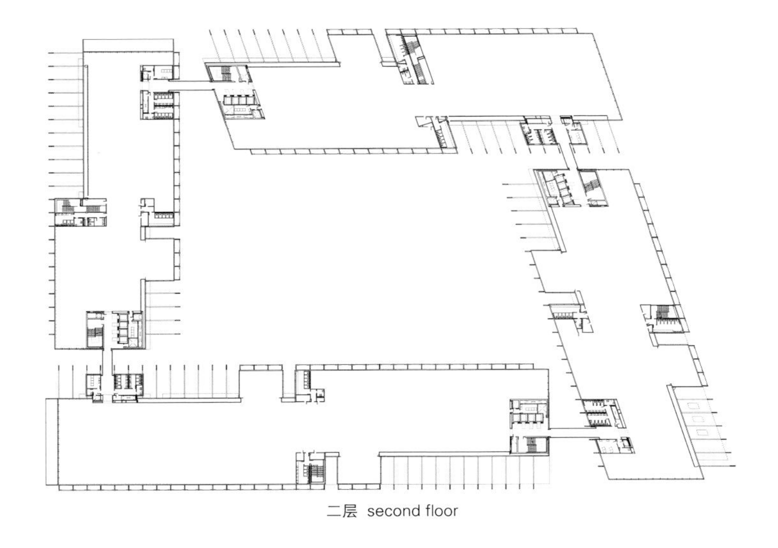

二层 second floor

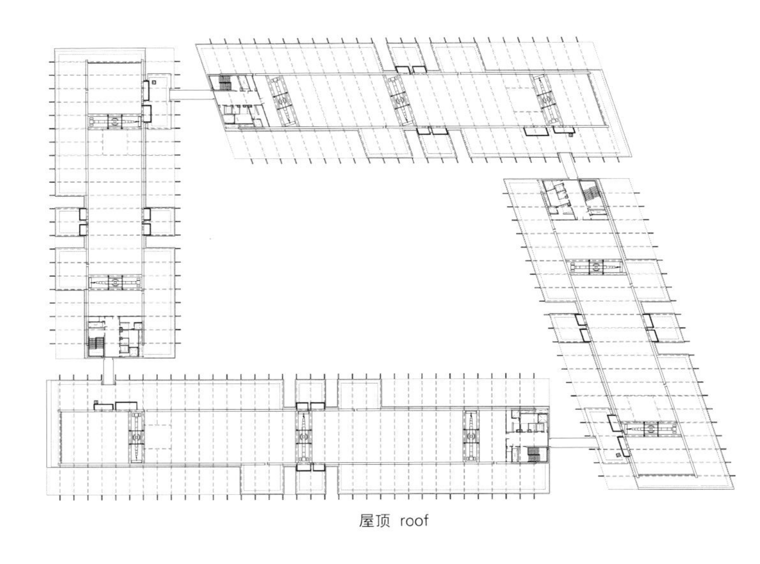

屋顶 roof

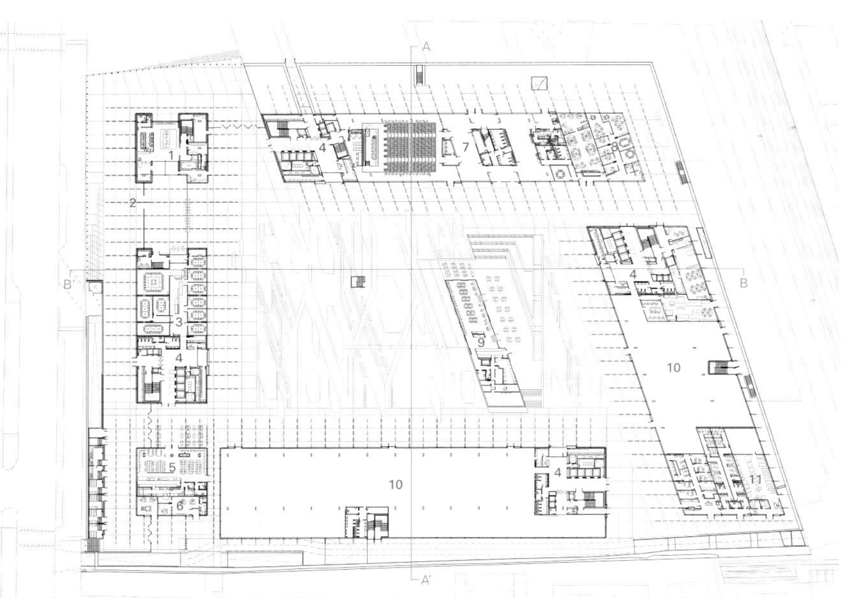

1 控制室&接待处 2 主入口 3 会议中心 4 中心区域 5 餐厅 6 银行
7 礼堂 8 工会办公室 9 自助餐厅 10 工作区 11 健康&运动中心
1. control&reception 2. main entrance 3. meeting center 4. core 5. restaurant 6. bank
7. auditorium 8. union office 9. cafeteria 10. working area 11. health&sport center

一层 first floor

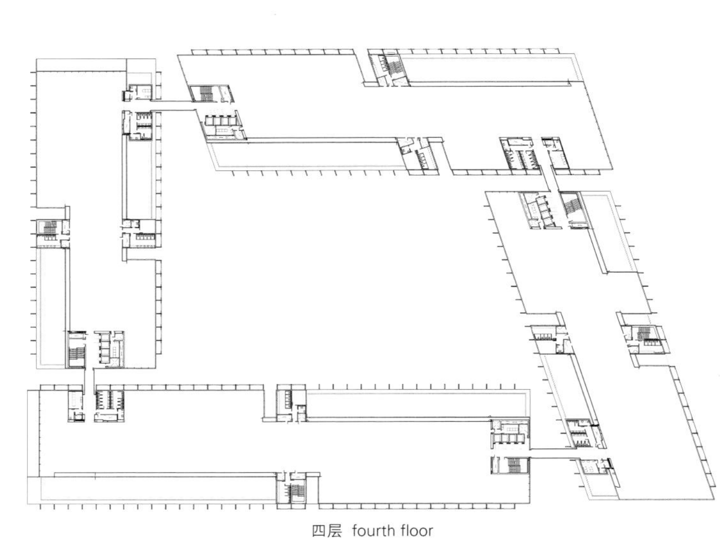

四层 fourth floor

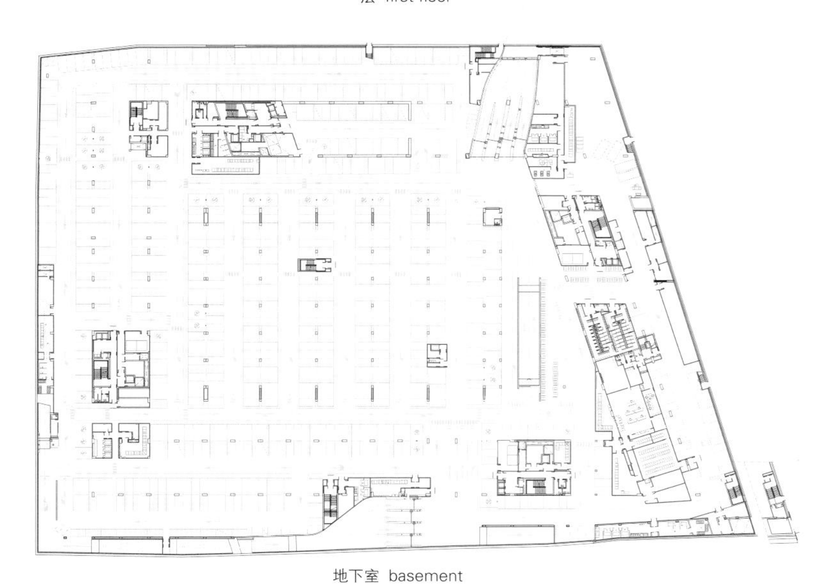

地下室 basement

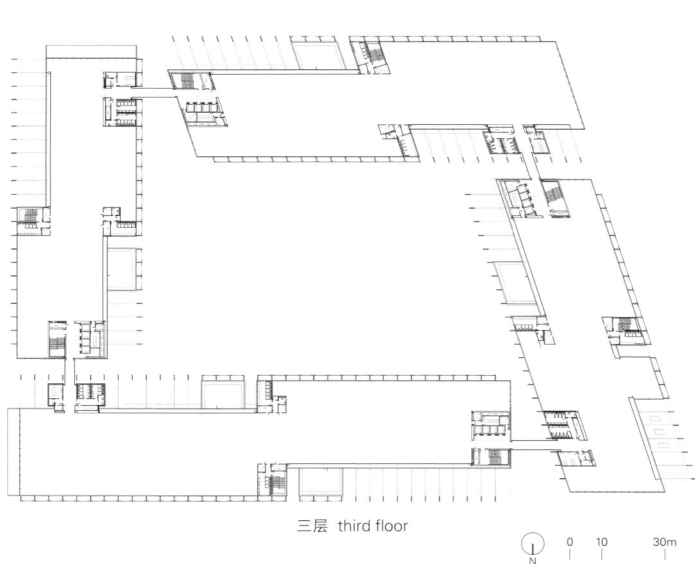

三层 third floor

the view on the decorative planting of the raised planting beds for over 4.000 employees. On the upper levels of the building, terraces shaded by climbers provide intimate outdoor spaces and improve the micro-climate during the heat of the Spanish summer. All elements are designed corresponding with the structure of the building: Large scale paving slabs, planting beds, water basin and the trees. Selection of plant species adapted to the climate, light reflecting materials, water management with cisterns and careful lighting helps to meet the high standards of LEED Certification for sustainability.

Sustainable Site

The chosen site enables the development of a former industrial area and existing infrastructure. It has an extensive network of public services (bus, subway, local rail, high-speed rail, etc.) and benefits from policies to encourage the use of bicycles, electric vehicles, high occupancy vehicles and low emission, fuel-efficient vehicles. A series of actions was also taken to encourage good relations with neighboring communities, such as watering the building site with non-drinkable water to avoid clouds of dust and cleaning the tires of lorries before they leave the site, pedestrianizing the area and providing the neighborhood with a telephone helpline during the building work.

Water Efficiency

Water consumption was optimized by designing the gardens to need less watering, choosing plant species suited to the Madrid's climate, using water-efficient equipment (bathrooms, taps, showers, etc.) and using rainwater for irrigation, which is stored in an underground tank with a capacity of 250,000 liters.

Energy and Atmosphere

Alternative energy generation methods were sought, such as the use of photovoltaic solar panels or gas heat pumps for climate control and hot water production.

Materials and Resources

From the design phase onwards, materials with a high-recycled content were chosen, reducing impacts from the extraction and processing of raw materials, supporting the use of local resources and encouraging responsible forestry.

Indoor Environmental Quality

As well as minimizing carbon emissions, users were encouraged to be particularly aware of a good living condition, by controlling lighting systems, maximizing interior natural light and providing pleasant views of the gardens. A comfortable temperature is ensured using the most advanced climate-control system that monitors the quality of breathable air.

Innovation in Design

A transport management plan was applied which quantifiably reduces the use of personal vehicles with multiple options and alternatives.

项目名称：Campus Repsol in Méndez Álvaro / 地点：C/Méndez Álvaro, 44, Madrid, Spain
建筑师：Rafael de La-Hoz Arquitectos
项目主管：Jesús Román, Concha Peña, Marcus Lassan, Carolina Fernández
首席设计师：Hugo Berenguer / 项目团队：Conchi Cobo, Guillermo Vidal, Bodo Schumacher, Laura Díaz, Hubert Lionnez, María Millán, Guillermo Cervantes, Beatriz Heras, Zaloa Mayor, María Martínez, Miguel García, Irene Zurdo, María Iglesias
施工单位：SACYR S.A.U. / 结构工程师：NB 35 / 安装工程师：R. Úrculo
景观建筑师：Latz+Partner, Javier Monge, Felix Metzler, Dörte Dannemann, Daniela Strasinsky, Thibault Le Marie / 室内设计师：Elena López de Meneses, Loreto Muñoz-Aycuens, Adriana Rodríguez, ANER / 工料测量师：Alberto González, Amaya Díaz de Cerio, Mercedes Esteban, José Luis Gonzalo
绘图：Luis Muñoz, Daniel Roris / 制模：Fernando Mont, Víctor Coronel
甲方：Repsol
用地面积：32,000m² / 总建筑面积：21,000m² / 有效楼层面积：123,400m²
设计时间：2009 / 竣工时间：2012
摄影师：©Alfonso Quiroga(courtesy of the architect)

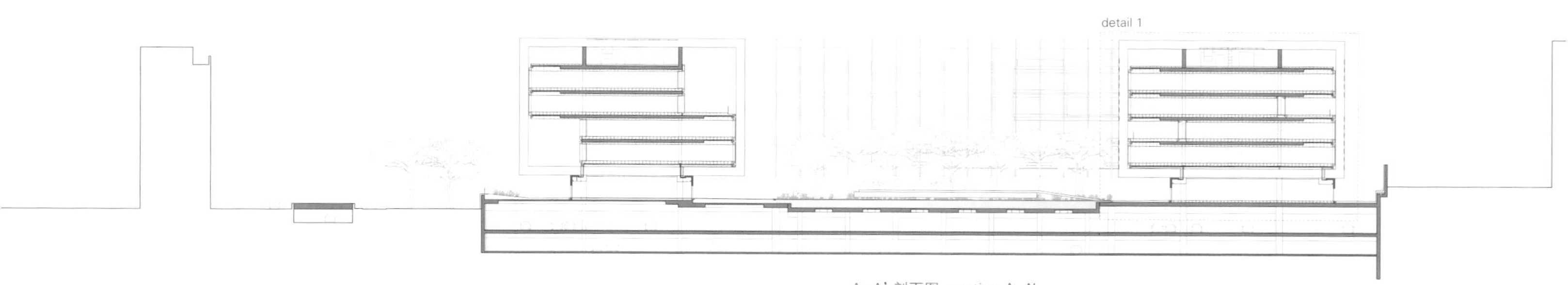

A–A' 剖面图 section A-A'

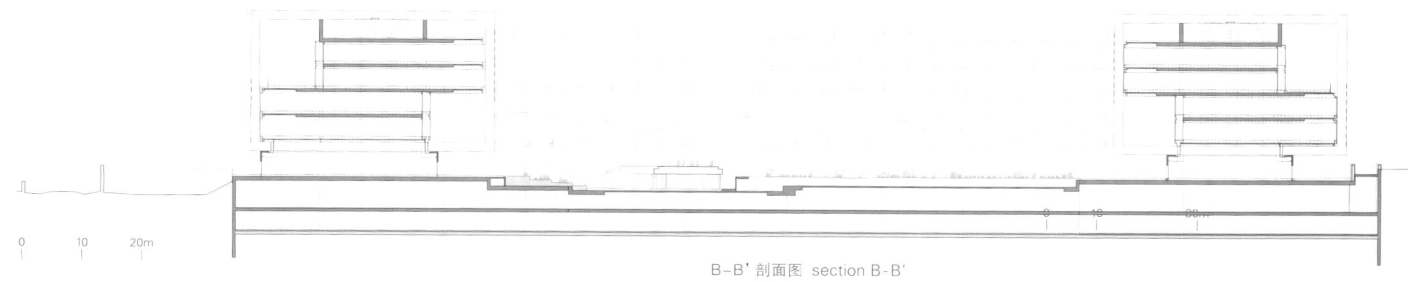

B-B' 剖面图 section B-B'

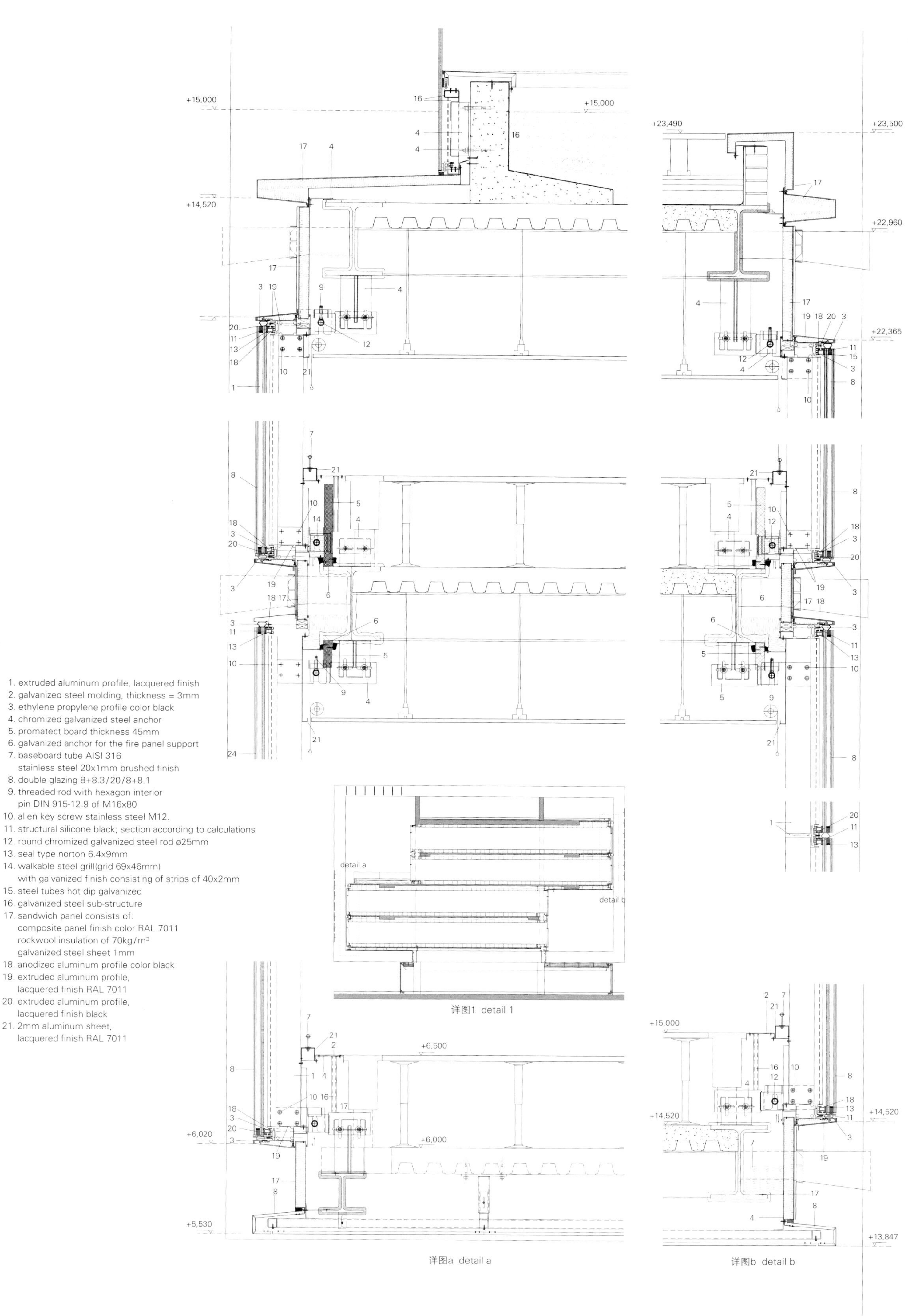

详图1 detail 1

详图a detail a

详图b detail b

亚利桑那州立大学的理工学院教学楼

Lake Flato Architects + RSP Architects

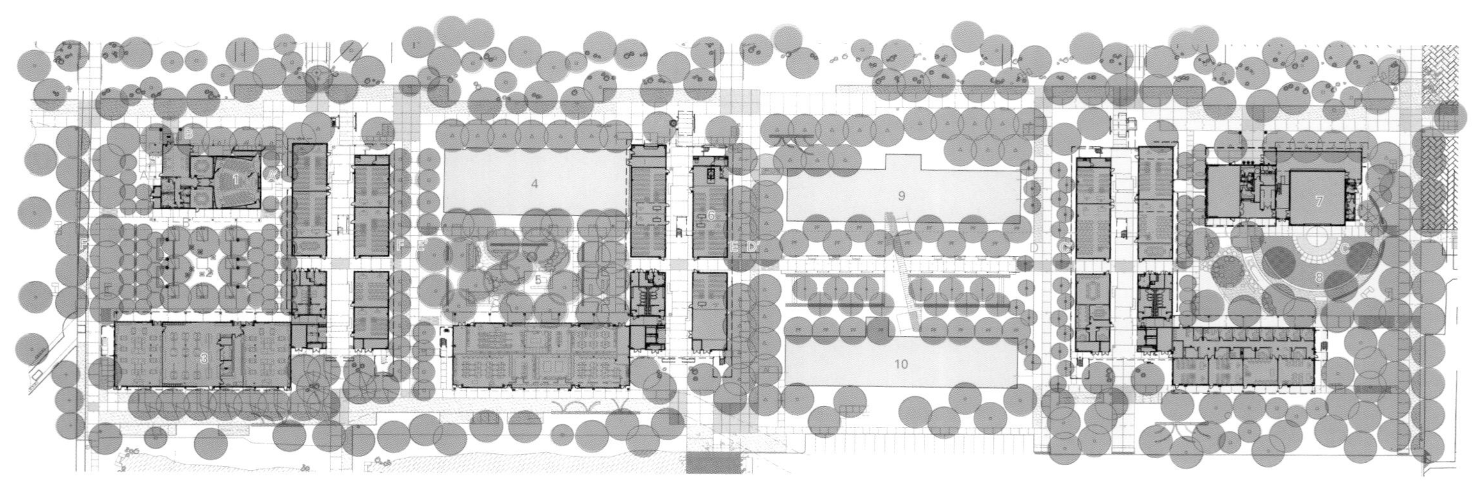

1 皮切卡乔大厅 2 农商庭院 3 佩拉尔塔大厅 4 Aravaipa礼堂 5 技术庭院 6 桑坦大厅
7 应用美术学院的展馆 8 艺术庭院 9 萨顿大厅 10 万纳大厅

1. Pichacho Hall 2. agribusiness courtyard 3. Peralta Hall 4. Aravaipa Auditorium 5. technology courtyard
6. Santan Hall 7. applied arts pavilion 8. arts courtyard 9. Sutton Hall 10. Wanner Hall

一层 first floor

一个全新的跨领域学术项目把曾经的空军基地建成了具有LEED金奖认证的、行人导向型的校园，使之成为位于美国亚利桑那州梅萨的亚利桑那州立大学的理工学院的一个全新的标志。这五栋全新的学术大楼容纳了四所完全不同却又相互联系的学院，主要设有科技和创新、管理和农商、应用美术和科学、教育信息以及师资培养这些学科。一座可以容纳500人的大礼堂为这四个学院服务。加上现存的三座大楼，共同形成一个连续的整体。这些建筑的四周被庭院环绕，各栋大楼通过一系列的露天中庭、建筑大门以及走廊连接起来。

这座建筑的设计反映出了理工院校的实践哲学，这点可以从大楼体系以及对特殊气候和环境做出的反应体现出来。修建这栋大楼所用的实用方法、其直接且重复的布局，以及对景观融为一体的强调，都使这座全新的"行人友善型"校园形成一个协同、学术型以及充满生机的社区。这个校园的中心强调了索罗兰沙漠栖息地。这条沙漠内的林荫路有许多天然的植被，因此形成了一条林荫大道。通过一系列的沟渠以及灌溉水渠，暴雨在场地进行了处理。从庭院屋顶所收集的雨水流到了沙漠内的林荫路中，在那里有许多小池塘，减缓了水流速度，且浇灌了这处天然的栖息地。

这五栋大楼都有着同样的叠加结构，综合教室都设在一、二楼，这样的设计安排有助于教室的调换，使额外电梯的使用需求最小化，也使学生能够穿过中庭到不同的地方。员工办公室基本都设在三楼，在这里可以看见北边山上的风景。实验室位于东西两翼结构内。

露天中庭取代了传统的内廊，增强了可见性并且突出了社区的感觉。这些露天中庭为大多数的教室提供了间接采光，平衡了从大楼边界进入多孔波纹遮光板的太阳光线。荫蔽且光线良好的空间设有阳台、狭小通道以及门廊，都是社交的理想区域，同时也减小了空调区域，并且将能量负载减低了15%。中庭里面的垂直机械管道使内部依旧保持原样，也有助于以后的整修或者重新规划。为了推进这个跨领域社区，中庭、共享会议室、荫蔽的庭院以及行人流线都为有组织或者是无组织的活动提供了契机。沙漠林荫路南面的三栋会议大楼标出了进入中庭的行人入口。一个高架走廊提供了社交或者是学习的空间，同时在共享会议室也能欣赏到沙漠林荫路和山的整体美景。中庭里面的木覆层凹室可以用作会议室、休息室或者是可供员工、老师、学生使用的大厅。这栋大楼的布局形成了园林庭院，这样一来最大限度地增大了荫蔽的程度。每个庭院都有一个独特的布局，为周边的空间增添互补的功能。比如，科技与人文庭院里面有一个沙漠沼泽地，可以收集并且过滤来自旁边场地或者屋顶的雨水，同时还设有户外教室和机械实验区域。农商学院的庭院种有果树丛和坚果树丛，它们都是当地的农作物，都为商业社区的社会活动提供了荫蔽。

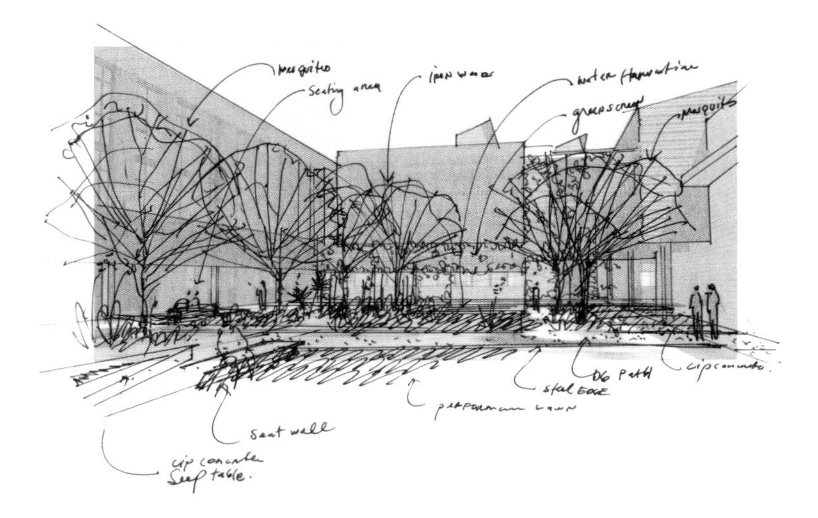

ASU Polytechnic Academic Buildings

A new interdisciplinary academic program transformed this former Air Force base into a LEED Gold-certified, pedestrian-oriented campus, and created a new identity for ASU Polytechnic in Mesa, Arizona. Five new academic buildings house four distinct yet interrelated colleges focusing on Technology and Innovation, Management and Agribusiness, Applied Arts and Sciences, and Education Innovation & Teacher Preparation. A 500-seat auditorium serves all four colleges. Integrated with three existing buildings to create a cohesive complex, the buildings are configured around courtyards and linked by a series of open-air atriums, building portals and arcades.

The buildings' designs reflect the Polytechnic Campus's practical philosophy through the expression of their systems and response to their particular climate and context. Through its practical approach to the architecture, straightforward and repetitive con-

7171

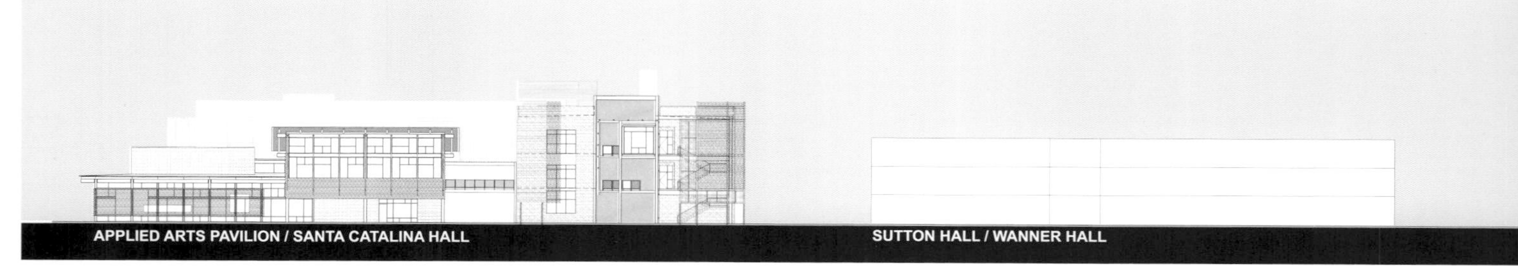
APPLIED ARTS PAVILION / SANTA CATALINA HALL
SUTTON HALL / WANNER HALL

功能示意图_桑坦大厅
program diagram_Santan Hall

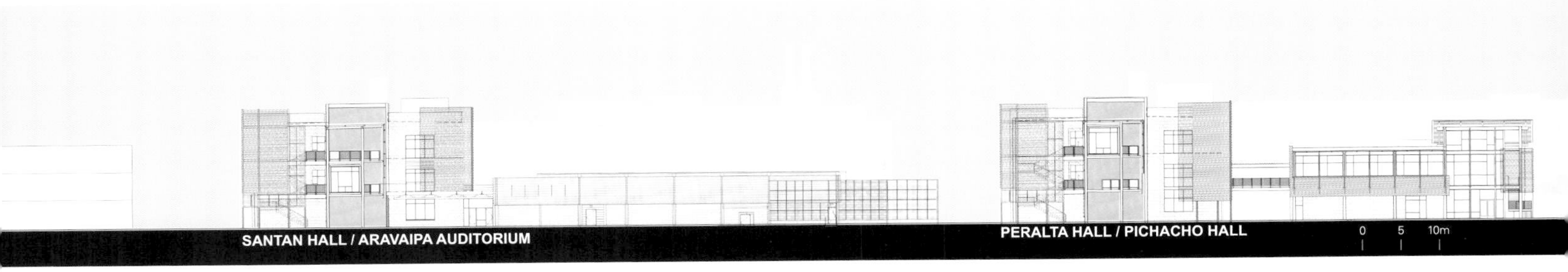

项目名称：ASU Polytechnic Academic Buildings
地点：Mesa, Arizona, USA
建筑师：Lake Flato
项目团队：principal in charge_Andrew Herdeg, lead designer_Ted Flato, sustainability coordinator_pm, Heather Holdridge, project manager_Matt Wallace
合作建筑师：RSP Architects
结构工程师：Paragon Structural Design
MEP：Energy Systems Design
LEED顾问：Green Ideas
土木工程师：Wood Patel & Associates
实验室顾问：Research Facilities Design
景观建筑师：Ten Eyck Landscape Architects
用地面积：56,656m²
有效楼层面积：22,761m²
施工开始时间：2006.11
竣工时间：2008.6
摄影师：©Bill Timmerman (courtesy of the architect)

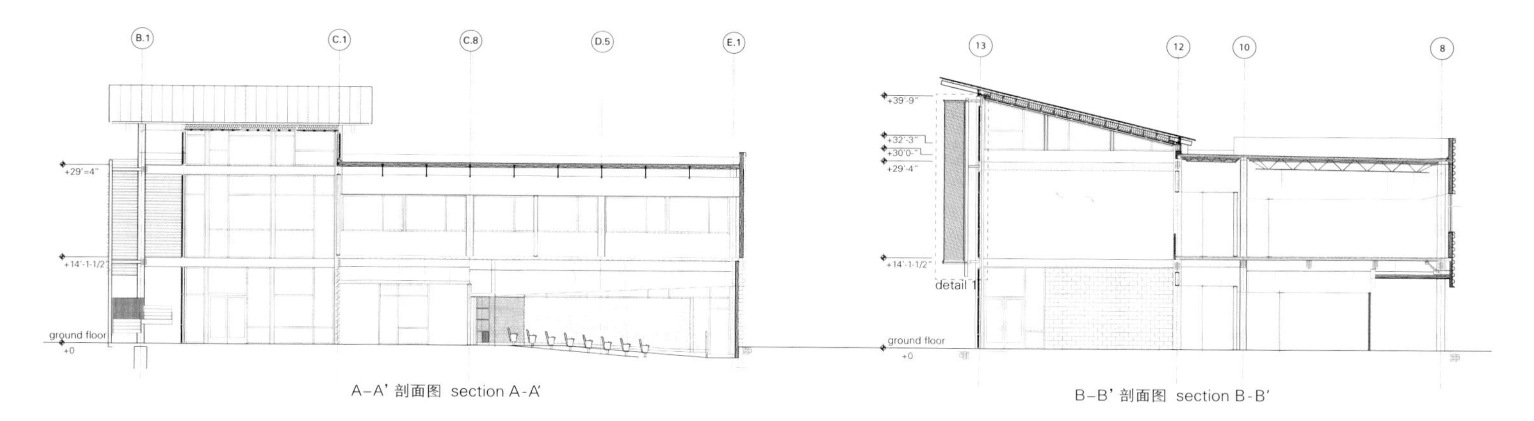

A–A' 剖面图 section A-A'

B–B' 剖面图 section B-B'

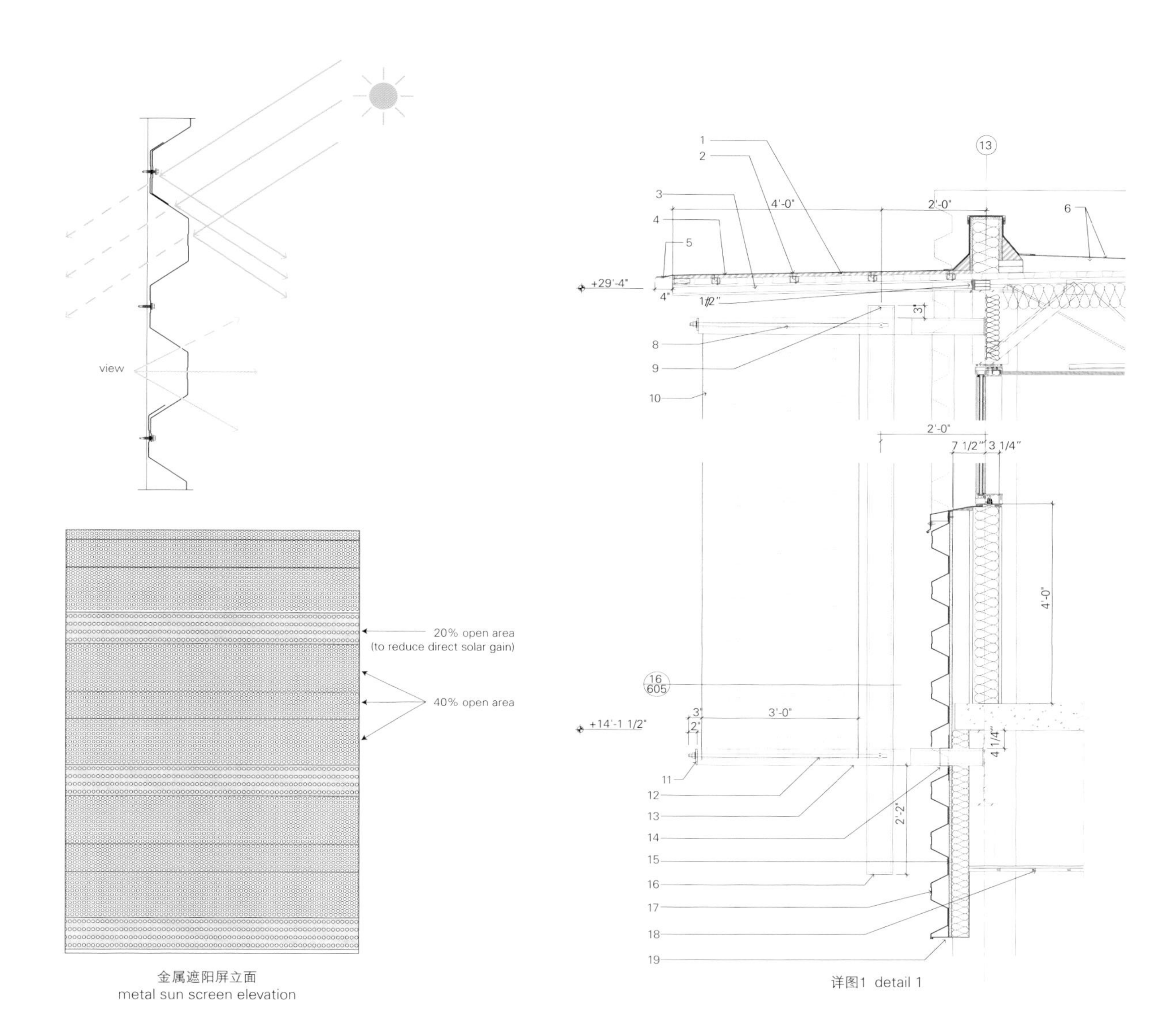

金属遮阳屏立面
metal sun screen elevation

详图1 detail 1

1. plywood sheathing
2. no exposed fasteners at exposed eaves of wood blocking inserted into valley of metal deck to provide anchorage from below the metal deck for plywood sheathing from above fasteners exceed 1-1/4" in length
3. paired channels see structural
4. membrane roof
5. bent plate
6. membrane roof on tapered rigid insulation
7. painted brake metal panel w/1/2" masonite backup panel coped to match paired channel profile
8. 3/4" steel rod welded to 4x6 HSS
9. steel cap to close top of tube steel
10. shade banner
11. 3/8"x5 1/2" x3" plt.
12. 3/4" steel rod
13. 4x 1/4"
14. break metal flashing around 4x4 HSS
15. 5/8" densglass sheathing
16. provide weep holes bottom of 4x6 HSS
17. 4" solid corrugated metal panel(morin O24W)
18. western red cedar wood soffit
19. breake MTL. flashing w. continuous sealant @edge slope 1.8" per foot @sill

figuration, and concentration on landscape integration, this new pedestrian-friendly campus fosters a collaborative, academic and vibrant social community. The heart of the campus celebrates Sonoran Desert habitat. The Desert Mall provides shaded walkways under native species. Stormwater is managed on site through a series of runnels and acequias. Rainwater is gathered from the roofs within the courtyards and conveyed to the Desert Mall where a series of small retention ponds slows the flow and is used to nourish the native habitat.

The five buildings share a common stacking diagram that places general classrooms on the first two floors facilitating access during class change, minimizing the need for additional elevators and exposing students passing through the atria to diverse programs. Faculty offices located primarily on the third floor take advantage of northerly views to the mountains. Labs are stacked in the east-west wings.

Open atria replace traditional corridors, increase visibility and foster a sense of community. They provide indirect daylighting to most classrooms, balancing daylight entering from the perimeter of the building through perforated corrugated sun screens. The shady, well-lit spaces house balconies, catwalks and porches ideal for socializing while reducing conditioned area and energy loads 15%. Vertical mechanical chases are located within the atria allowing the interiors to remain unobstructed and more able to be reconfigured for future growth or reprogramming. To promote an interdisciplinary community, atria, shared conference rooms, shady courtyards, and pedestrian circulation provide opportunities for both structured and unstructured interactions. Three conference towers on the south side of the Desert Mall mark the pedestrian entries into the atria. An elevated porch provides areas to socialize or study, while shared conference rooms provide views up and down the Desert Mall and views to the mountains. Wood clad bays within the atria serve as conference rooms, break rooms, or lobbies for faculty, staff, and students. The buildings are configured to form landscaped courtyards proportioned to maximize shade. Each courtyard has a unique configuration providing complimentary functions for surrounding program spaces. For instance, the science and humanities courtyard houses a desert wetland that captures and filters stormwater from adjacent site and roofs, and includes an outdoor classroom and mechanical lab area. The courtyard at the School of Agribusiness uses a bosque of fruit and nut trees, the local agriculture crop, to provide shade for social events with business community.

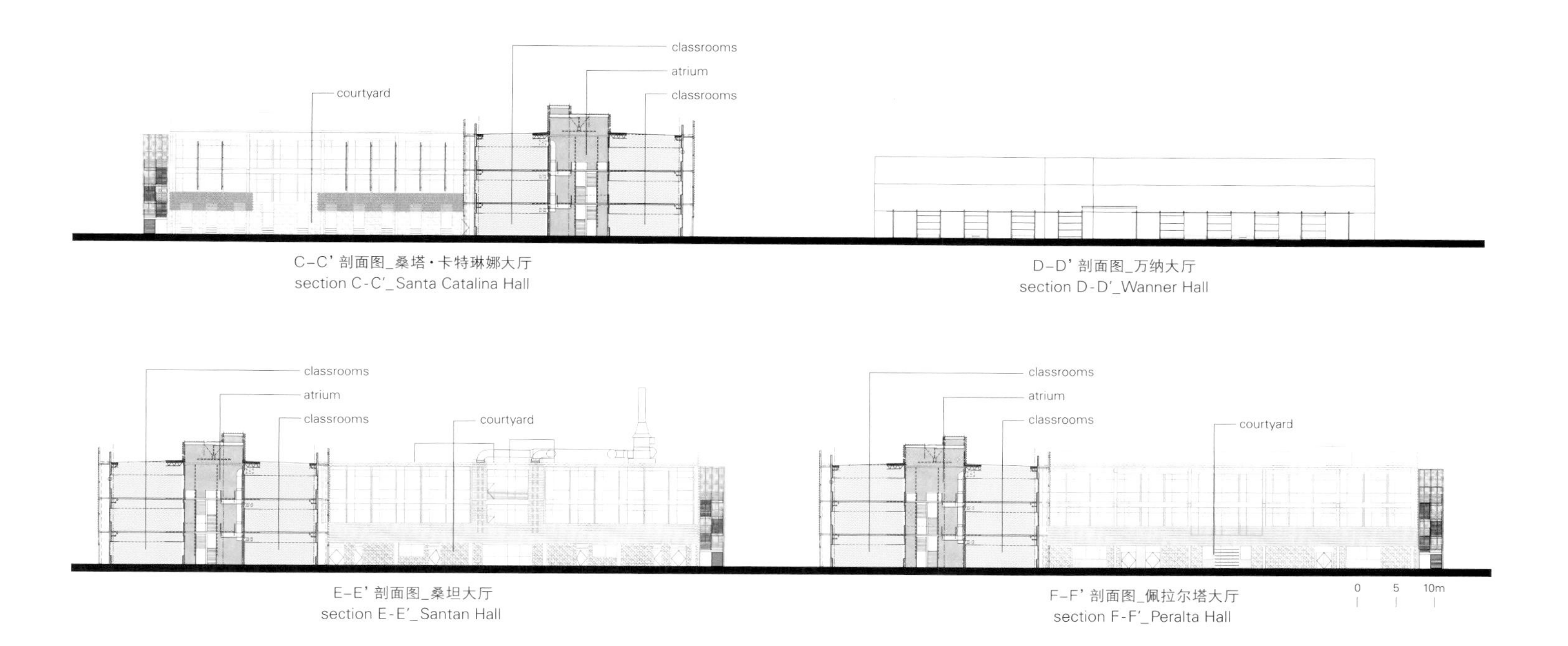
courtyard
classrooms
atrium
classrooms
C-C' 剖面图_桑塔·卡特琳娜大厅
section C-C'_Santa Catalina Hall
D-D' 剖面图_万纳大厅
section D-D'_Wanner Hall
classrooms
atrium
classrooms
courtyard
E-E' 剖面图_桑坦大厅
section E-E'_Santan Hall
classrooms
atrium
classrooms
courtyard
F-F' 剖面图_佩拉尔塔大厅
section F-F'_Peralta Hall
0
5
10m

Fundamentals

14th Venice Biennale

基本法则

第十四届威尼斯双年展概况

2014年威尼斯建筑双年展在万众期待中揭开了帷幕，这主要归功于其策展人的选择，即荷兰建筑师雷姆·库哈斯：展览在揭晓之后，将游客分为两部分，一部分人对所有展览的造价表现出极度的兴奋，而另一部分人则是消极地对这些展览进行挑剔。

在威尼斯展的历史长河中，第一次需要展出策展人所做的个人研究，与之前的展览进行对比，并且试图通过一系列受邀建筑师和展览的建筑来估测当代建筑的趋势。

而今年，双年展试图定义建筑的全新起点，并且对基本法则进行分析，这些法则将从不同的角度对建筑进行调查，且对所有的建筑设计通用，无关地点、设计师和时间。

最终产生了一个灵活的展览，信息丰富，人们必须进一步探索来进行分析，但同时缺乏建筑最重要的元素：空间。

The opening of the Biennale Architecture in Venice in 2014 has been preceded by a very high expectation, because of the choice of the curator, the Dutch architect Rem Koolhaas: a choice that then, after the opening of the exhibition, divided the visitors between the ones excited at all costs and the negative hypercritical ones.
For the first time, in the course of its long history, the Venetian exhibition indeed had to deal with the exhibition of a personal research made by the curator, in contrast to what happened in the previous editions, which tried to take stock of the situation of contemporary architecture through the construction of long lists of invited architects and shown architectures.
This year the Biennale tries to define a sort of new starting point for architecture, proposing the analysis of Fundamentals, which investigate the architecture itself from different points of view, that should be common to all architectural design, regardless of location, designer, and time.
The result is a little accommodating show, very dense of information, difficult to analyze in a superficial way and, simultaneously, rather poor of the most important element of architecture: the space.

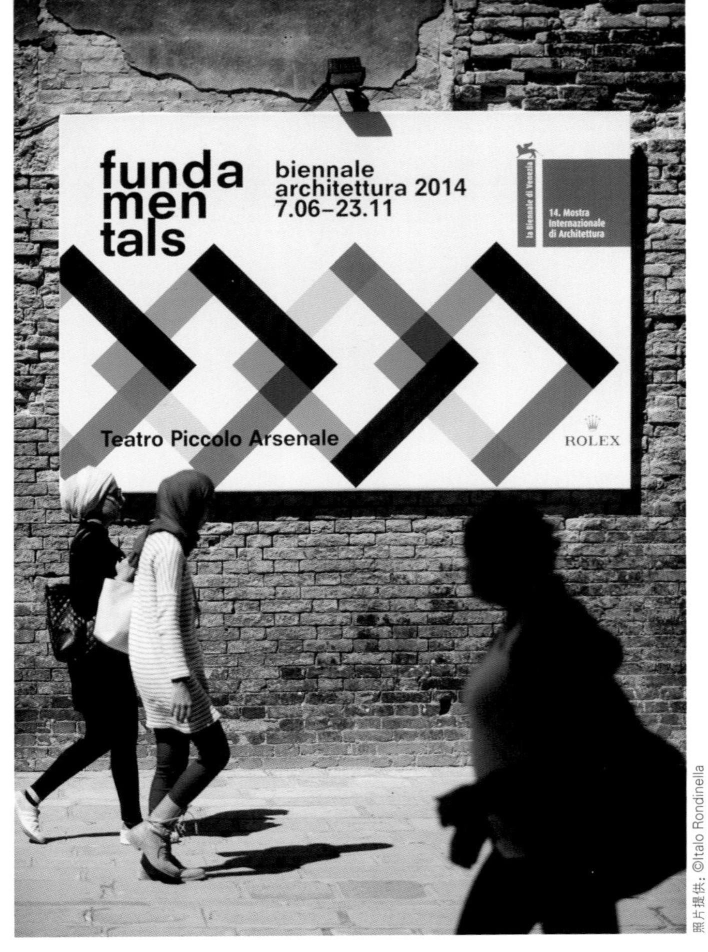

照片提供：©Italo Rondinella

照片提供：©Giorgio Zucchiatti

“建筑的元素”展览入口，中央展亭
Elements of Architecture entrance, Central Pavilion

照片提供：©Gilbert McCarragher

OMA与施华洛世奇合作制作的灯具，体现在军械库的Monditalia展馆中
Luminaire by OMA in collaboration with Swarovski, Monditalia exhibition in Arsenale

照片提供：©Giorgio Zucchiatti

根据圣马修、由Virgilio Sieni制作的福音书，位于Monditalia展馆的舞区
Gospel according to Saint Matthew by Virgilio Sieni, The Dance sector in the Monditalia section

雷姆来到了威尼斯

OMA，或者关于叙述性目录

当人们浏览雷姆·库哈斯于1995年设计的大书（OMA作品选集），《S, M, L, XL》时，便会注意到在Dall' Ava别墅项目的页码中，介绍了两个重要的元素：照片（被束缚的长颈鹿，在房屋的室外空间移动）的超现实表现，以及一个不同寻常的图纸设计方式（还在进行中，满布手写的红色注释，展示了变化、评论以及替代性）。

在这两种元素中，人们可能辨别出典型的荷兰事务所设计的项目：这类项目的模式是强调过程，而这个过程似乎是将建筑本身的重要性最小化，并且在关注的其他方面，又强烈地回归到展示的对象本身。Dall' Ava别墅项目主要特点的展示要归功于其叙述体系，这个体系通过将其特点从大众背景中释放出来，来对其进行强调：如果我在花园中看见一头长颈鹿，毫无疑问，我将会更近距离地观看它所处的场地，如果我带着好奇心来解读这些较为常见的信息，那么我将会更加关注这些细节（对项目所做的说明）及项目本身。

然而，这一进程要被适量的叙述元素所支撑，从不同的角度来说，是因为它属于可能出现的建筑中，如荷兰工作室预期的那样。之后，这一叙述的焦点才是对元素目录的定义，这一目录定义了多样性视点，而这种视点偏离了建成空间。

《Content》杂志一书于2004年开始出版，并且继续收录OMA的作品，是理解这种方法的理想纲要，而其也在介绍这个作品的简要对话中进行了清晰的定义：

“我不确定这是一本书，还是一本杂志。”

Rem goes to Venice

OMA, or about the narrative catalog

When browsing the pages of *S, M, L, XL*, the brick-book presented by Rem Koolhaas in 1995 as an anthology of OMA's work, one may notice, in the pages dedicated to the Villa Dall'Ava project, two significant elements: a surreal appearance of the photograph (a giraffe on a leash, moving in the house's outdoor spaces) and an unusual approach to the design drawings (in progress, full of handwritten notes with a red pen, indicating changes, comments, alternatives).

In these two images it is possible to discern a means of showing architecture typical of the Dutch office: it is a mode that emphasizes a process, which seems to minimize the importance of the architecture itself, but which, in its focus on something else, returns strongly to the object shown. The main features of the Villa Dall'Ava emerge thanks to the system of narration that emphasizes the architecture's characteristics by freeing them from a generic background: If I see a giraffe in a garden, I will definitely tend to look more closely at the place where it is located, and if I read, with some curiosity, "informal" notes, my approach to the details on which these notes comment and, ultimately, to the project itself, will be more careful.

Necessarily, however, this process is supported by an adequate number of narrative elements, from different points of view, because it is in this quantity that the architecture may emerge, as intended by the Dutch studio. The focal point of this narration, then, is the definition of a catalog of elements – a catalog that defines a multiplicity of points of view, tangential to the built space.

Content, the magazine-book published in 2004, which continues to collect OMA projects, is an ideal compendium for understanding this approach, and is clearly defined in the brief dialogue that introduces the work:

"I'm not sure if this is a book or a magazine."

特殊活动
Collateral Events

1. Across Chinese Cities–Beijing
2. Adaptation
3. Air Fundamental. Collision between inflatable and architecture
4. Fundamentally Hong Kong? Delta Four 1984–2044
5. Gotthard Landscape: The Unexpected View
6. Grafting Architecture. Catalonia at Venice
7. "Happiness Forecourt" = "Largo da Felicidade" = "开心前地"
8. Lifting The Curtain: Central European Architectural Networks
9. M9/Transforming the City
10. Made in Europe
11. Masegni
12. Mikhail Roginsky. Through the Red Door
13. Moskva: urban space
14. Once upon a time in Liechtenstein
15. Planta
16. Salon Suisse: The next 100 Years–Scenarios for an Alpine City State
17. The Space That Remains: Yao Jui-Chung's Ruins Series
18. The Yenikapi Project
19. Time Space Existence
20. Township of Domestic Parts: Made in Taiwan
21. Young Architects in Africa
22. Z Club. On Money, Space, Postindustrialization, And...

Monditalia展馆
Monditalia

1. The Room of Peace 2. Italian Ghosts
3. Post-Frontier 4. Intermundia
5. Theaters of Democracy
6. The Third Island Ag '64 '94 '14
7. The Architecture of Hedonism-Three Villas on The Island of Capri
8. Legible Pompeii
9. Pompeii, The Secret Museum and the Sexopolitical Foundations of The Modern European Metropolis
10. Antonioni's Villa
11. La Maddalena
12. 99 Dom-ino
13. Rome–San Giacomo Hospital The Ghost Block of Giambattista Nolli
14. Cinecitta Occupata
15. A Minor History Within the Memories of A National Heritage
16. All Roads Lead to Rome. Yes, But Where Exactly?
17. L'aquila's Post-Quake Landscapes(2009-2014)
18. Assisi Laboratory
19. Space Electronic: Then and Now
20. Ground Floor Crisis
21. Superstudio. The Secret Life of the Continuous Monument
22. Biblioteca Laurenziana
23. The Remnants of the Miracle
24. Nightswimming: Discotheques in Italy From the 1960s until Now
25. Dancing Around Ghosts–Milano Marittima's Panem et Circenses
26. Urbs Oblivionalis
27. Urban Spaces and Terrorism in Italy
28. The Landscape Has No Rear
29. Tortona Stories
30. Countryside Worship
31. Radical Pedagogies: Action-Reaction-Interaction
32. Architecture of Fulfilment: a Night with a Logistic Worker
33. La Fine Del Mondo
34. The Business of People Effimero: Or the Postmodern Italian Condition Immediate Surroundings
35. Residences of Italian Mafia Organizations
36. Sales Oddity. Milano 2 and the Politics of Direct-To-Home TV Urbanism
37. Z! Zingonia, Mon Amour
38. Designing the Sacred
39. Italian Limes
40. Alps

吸收现代性
Absorbing Modernity

1. Republic Of Korea
2. France
3. Russia
4. Canada
5. Belgium
6. Serbia
7. Germany
8. Australia
9. Austria
10. Brazil
11. Denmark
12. Egypt
13. Finland
14. Japan
15. Great Britain
16. Greece
17. Israel
18. Netherlands
19. Nordic Countries
20. Venezia
21. Poland
22. Czech Republic
23. Romania
24. Spain
25. USA
26. Switzerland
27. Hungary
28. Uruguay
29. Croatia
30. Estonia
31. Indonesia
32. Ireland
33. Kuwait
34. Latvia
35. Malaysia
36. Morocco
37. Portugal
38. Kingdom Of Bahrain
39. Republic Of Kosovo
40. Republic Of Slovenia
41. Thailand
42. Chile
43. Dominican Republic
44. Mozambique
45. Albania
46. Argentina
47. Costa Rica
48. United Arab Emirates
49. Macedonia
50. Iran
51. Mexico
52. Peru
53. South Africa
54. Turkey

>> Monditalia展馆+吸收现代性_军械库
Monditalia + Absorbing Modernity_Arsenale

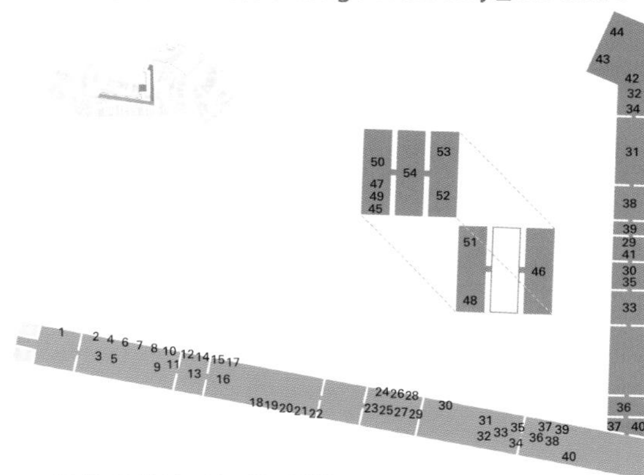

>> 吸收现代性_Giardini展馆
Absorbing Modernity_Giardini

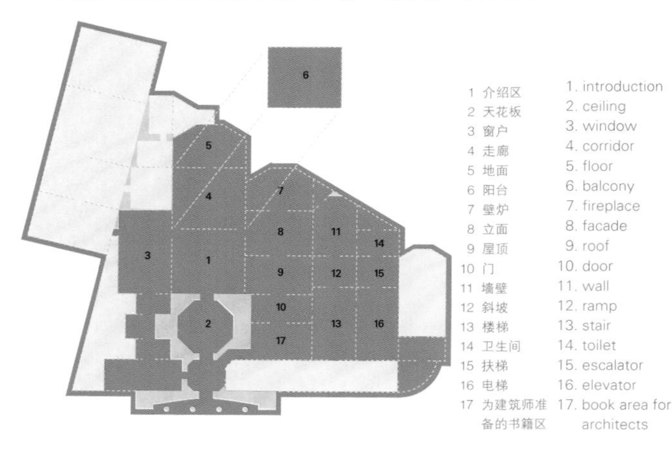

窗户

窗户过去常用来制造空间，它创造了场所性。靠窗座位、窗台、凸窗、游廊、百叶窗、窗帘都标记了窗户在立面上和房间内的位置。自20世纪以来，窗口和玻璃生产技术的进步也允许大多具有细微差别的本地组件安装在窗户结构内部，而人们无法看见它们。玻璃，最初似乎与窗户是绝佳组合，且在幕墙中完全成为主导，而幕墙是一种西方的发明，允许其他地区坚持建筑要求，并且从其历史话语中解放出来。

Windows

The window used to make space, it asserted placeness. Window seats, sills, bay windows, verandas, shutters, blinds, curtains all marked the position of the window on the facade and in the room. Since the twentieth century, technological advances in window profiles and glass production have allowed many of these nuanced local components to be internalized in the window's structures, magically invisible. Glass, which initially seemed the perfect partner for the window, took over entirely, culminating in the invention of the curtain wall – a Western invention, which allowed other regions to stake a claim in architecture, liberated from its historic discourses.

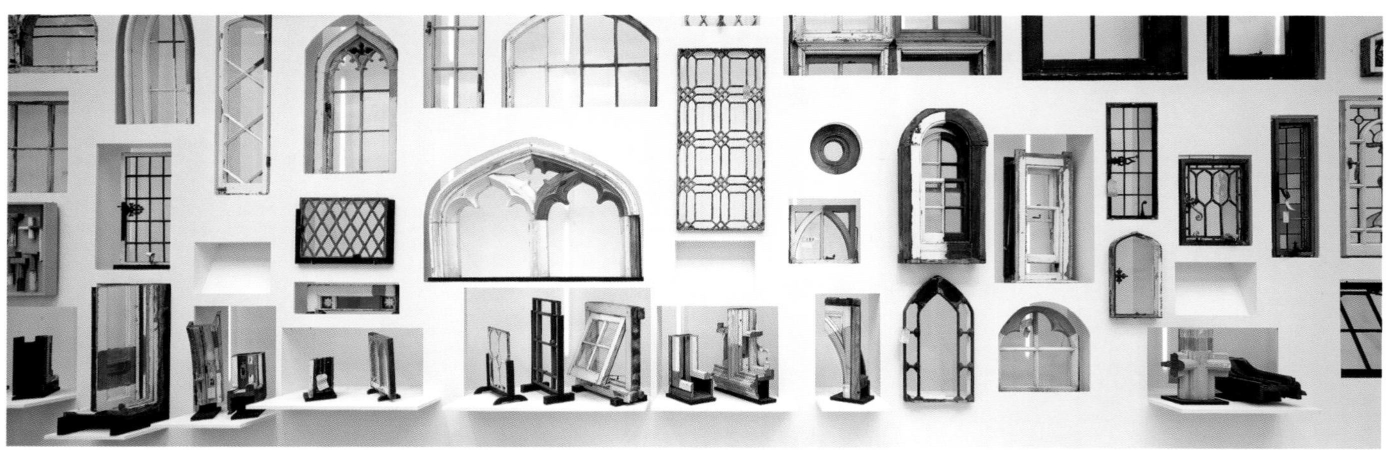

"事实上，我在这两种超级有趣的元素中寻找到了张力。"

这不是巧合，最后，2012年位于伦敦Brabican中心的、由策展人Belgian Rotor发起的OMA作品展览，则证明了在《S,M,L,XL》一书出版后的二十年里，荷兰工作室的叙述方式并没有发生太大的改变。

这次展览只是对一些四散的材料进行编录，且对其进行摘录。这些材料存于积累中，存于数量中，存于其合法的存在中。此外，不像考古学家会深挖进泥土底层，Rotor只是将游客分成若干组合，将他们与精心设置的间离效果并置在一起，来突出更多建筑的出现。

另一方面，OMA的作品通过其叙述性的积累，来发展自身，并且对现代主义、流行文化、公开展示的项目以及历史提议的引证进行总结。感谢雷姆·库哈斯，后现代建筑方法找到了一个实际的转折点，从怀旧主义中解脱出来，并且限制了其范围（没有涵盖整个建筑历史，而仅仅是现代历史），更重要的是，利用典型的超现实主义的讽刺性和异化性，来把生活带入到现代建筑的某些杰作中。

狂喜的威尼斯？

雷姆·库哈斯被要求来策划第十四届威尼斯建筑展，他也利用他的设计方式、社论以及教育方法来继续其工作模式。

威尼斯展是一次丰富的信息积累，是荷兰建筑师与哈佛大学合作两年的结果，并且展现在三个层次中：建筑的元素、Monditalia（指意大利

"Actually, I find the tension between the two super-interesting."

It is no coincidence, finally, that the exhibition of OMA's work curated by Belgian Rotor at the Barbican Centre in London in 2012, provided evidence that almost twenty years after the publication of *S, M, L, XL* the Dutch studio's narrative approach has not changed significantly.

The show is nothing but excerpts from a vast catalog of disparate materials, which are found in their accumulation, in their quantity, their raison d'être. And, unlike archeologists that dig deep into the soil's stratigraphy, Rotor did little more than bring visitors into a number of slices, juxtaposing them with a carefully cultivated alienating effect, to emphasize even more the emergence of the architecture.

On the other hand, it is via narrative accumulation that OMA's projects develop themselves to provide a fine summation of citations of modernism, pop culture, spectacular engineering, and historical suggestions. Thanks to Rem Koolhaas, the post-modern approach to architecture has found a real turning point, freed from nostalgia, limiting its range (covering not the entire history of architecture, only that of the modern), but, more importantly, using the irony and alienation typical of surrealism to breathe life into some of the masterpieces of contemporary architecture.

Delirious Venice?

Called upon to curate the XIV Biennale of Architecture in Venice, Rem Koolhaas has coherently continued the operating mode typical of his approach to design, editorializing, and education.

The Venice Exhibition is a copious accumulation of information, the result of two years of research carried out by the Dutch architect in collaboration with Harvard University, revealed in three "stratigraphic" slices: Elements of Architecture, Monditalia, and Absorbing Modernity: 1914-2014. These pieces of evidence do not require sequential congruence, given that they are juxtaposed, but they are distinguished by a common approach, as the sum-

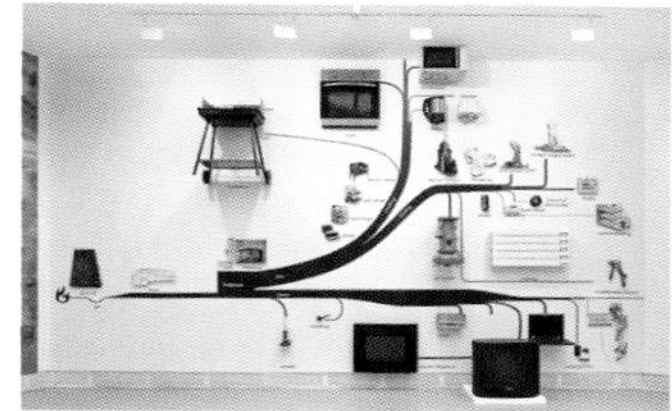

壁炉
技术赋予壁炉以活力，而如今，壁炉作为建筑内的一个具体的设备或者场地，基本上都消失了，而且我们几乎没有注意到这种情况。之前壁炉的功能是加热、烹饪、照明，并且是人们的聚集地，也是媒体和文化的焦点所在。而现在壁炉已经被分割，置于多样的设备之间，或者像贯穿于建筑系统的触手一样蔓延在建筑内。炉子可能成为第一个消失的建筑元素。

Fireplace
The promethean technology of the fireplace has now more or less disappeared as a discrete object or place within architecture, and we have hardly noticed this erasure. The former tasks of the fireplace – heating, cooking, lighting, a gathering place and focal point for media and culture – have been divided up among multiple devices, and/or spreaded like tentacles throughout various building systems. The hearth may become the first architectural element to become extinct.

地面
曾经建筑师应用了一种能够进行象征性表达的表面，它定义了空间的使用方式，名为“游戏的规则”。20世纪所应用的地面趋向于一种纯粹的笛卡尔曲面，设计合理，未进行装饰，呈现理想中的平坦状态，毫无杂音。同时，由于我们对其功能、象征以及触觉方面的潜在性的疏忽，地面空间成为建筑空间的主要经济隐喻，称之为“平方米主义”。但是在房地产用语中，平方米是一个贯穿整个空间的三维体量，而地面本身就是一块厚板，和天花板一样神秘。

Floor
Once a surface for symbolic expression – defining the way spaces are used, the "rules of the game" – floors in the twentieth century tended towards a purely Cartesian surface, rational, undecorated, always perfectly flat, ideally soundless. Simultaneously with our negligence of their programmatic, symbolic, and haptic potential, floorspace became the dominant economic metaphor for architectural space: call it square meterism. But the square meter, in the parlance of real estate, is really a 3-dimensional volume through the entire space. And the floor itself is actually a thick slab, sometimes a "false floor" containing mysteries similar to those of the ceiling.

宏图背景）以及吸收适应性：1914—2014。考虑到它们是并列的，因此这三个元素并未要求具有连贯的一致性，而是采用一个通用的方法将其区分开来，作为所参考的高密度数据的总结。

基础建筑

“建筑的元素”是这次展览的主题，主要集中在策展人的直接陈述以及大量的评论、调查和批判中，它们或者是非常积极赞成，或者是处于开放的反对状态。建筑的元素首先是一个目录，附在15本书中的内页中，之后被抽出，且部分通过空间发展来进行展示。

库哈斯好像有意突出与之前展览的距离，并且已决定通过一个建筑师的摹写单和建筑来诠释其策展人的职务，并且试图对空间实质进行调研，通过15种观念来对其进行仔细的分析，仿佛要重新突出从大量信息中分层次抽取的主题。

根据荷兰建筑师的看法，这15种观念是定义空间及其存在的基本元素，不受设计建筑和场地的限制，其中包括历史和经济条件。这便是2014双年展的优势：试图将建筑演讲从原创者的独白、时尚所带来的影响、外界参考以及自我参考的职业论述中释放出来。在这里，库哈斯试图展示设计师之间的普通对话的要点，尽管具有间接性，其要点也要比之前的由大卫·科波菲尔策展的双年展（主题为共同点）要多，且能够清晰地体现出来。

ming of data from a high density of references.

Elementary Architecture

Elements of Architecture is the exhibition's main subject, on which is concentrated the curator's direct narration and around which has accumulated a huge number of comments, investigations, criticism, largely positive or openly hostile.

Elements of Architecture is first a catalog, enclosed in the pages of fifteen books, that is then extracted and shown in part through spatial development.

As if to highlight the distance from previous exhibitions, Koolhaas has decided to emancipate his curatorship through a slavish list of architects and architectures, trying to investigate the very essence of space, dissecting it through fifteen points of view, as if to reemphasize the theme of stratigraphic extraction from a vast quantity of information.

The fifteen points of view are, according to the Dutch architect, the basic components that define the space and its presence, independently of those who design the building and the place in which it is located, including its history and economic conditions. Such is the strength of the Biennale of 2014: trying to free architectural discourse from the sole claim of authorship, the influences of fashion, the external references, the self-referential discourse of the profession. Koolhaas tries here to suggest real points of common dialogue between designers, even if indirectly, more so than did the previous Biennale, curated by David Chipperfield (Common Ground), which explicitly sought them.

The exhibit can, therefore, provide a strong rupture point as compared to previous shows, becoming a fundamental starting point for the future. Asked to curate the exhibition, Koolhaas, after reportedly researching the project for two years, insisted upon a longer showing, the six months exhibition accorded the Biennale of Art, instead of architecture's usual three, finally giving the same importance to the two exhibitions.

照片提供：©Giorgio Zucchiatti

门
一旦一个传统元素被赋予了物理分量和图形图像，它便会转为一处非物质区域，成为临时使用的技术（而非实际物体）所定义的环境之间的过渡，这种转换与社会的转型同时发生：隔离是曾经所需的条件，而现在我们的愿望就是为移动、流动、透明性、可达性服务，而这些都受到了门的阻碍。

Door
A traditional element once invested with physical heft and graphic iconography has turned into a dematerialized zone, a gradual transition between conditions registered by ephemeral technologies rather than physical objects. The transformation took place concurrently with a transformation in society: whereas isolation was once the desired condition, our aspirations now are for movement, flow, transparency, accessibility – which the door, by definition, stands in the way of.

卫生间
没有建筑论文将卫生间引用为原始的建筑元素，但是今天卫生间成为人类与建筑在最亲密的范畴内发生互动的基本区域，曾经在罗马的城市有一项受人尊敬的社交活动便是去卫生间，这一行为逐渐变得私有化，在建筑内的封闭区域进行。在19世纪，冲刷厕所的技术、S形存水弯和现代化管道，使卫生间与浴室设在一个单独的房间，脏乱与整洁的结合，在历史上被成功地实现了若干次。卫生间的本地化、私有化以及扩散性是许多建筑和城市规划背后的无言的推动力。卫生间曾经是最私密的也是最政治化的元素，自从1539年弗朗索瓦国王指令巴黎市民负责收集和妥善处理他们的水源以来，便受到了政府的干预。而今天，卫生间成为文化叠加、抵抗力以及进行慈善事业的场地（比尔和梅林达·盖茨基金会所面临的“改造厕所”的挑战），也是看起来很棘手的习惯的发生地。

Toilet
No architectural treatise cites the toilet as the primordial element of architecture, but the toilet is today the fundamental zone of interaction between humans and architecture on the most intimate level. Once a respectable communal activity in Roman cities, going to the toilet gradually became privatized, enclosed within architecture. In the nineteenth century, enabled by flush technology, the S-trap, and modern plumbing, the toilet united in a single room with the bath – a union of the dirty and the clean that had only been safely achieved a handful of times in history. The domestication, privatization, and proliferation of the toilet are the great unspoken driver behind many architecture and urban planning. The toilet is at once the most private and the most political element, subject to government interference at least since King Francois' 1539 edict instructed the citizens of Paris to take responsibility for the collection and proper disposal of their "waters." Today, the toilet is the site of cultural superimpositions and resistance, philanthropy (Bill and Melinda Gates Foundation's challenge to "reinvent the toilet"), and habits that only seem to be intractable.

因此，和之前的展览相比，这个展览能够提供一个裂点，使之成为开拓未来的起点。自库哈斯被邀请策展这个展览以来，他用两年的时间对项目进行调研，并坚持举办一个更长时间的展览，根据艺术双年展的情况，确定为期6个月，来取代建筑展通常的三个月期限，并且其重要性与两个展览同等。

为了与进行的Massimiliano Giono策展的艺术双年展（主题为百科殿堂）保持一致，库哈斯采用了纪录片的方式，提供了一个案例研究的百科全书，来反映建筑体以及在空间形式内得到解释的，使人回想起19世纪建筑手册的参考性案例。在这些解释中，我们看到了策展人的选择限制。精选的构件种类十分丰富，以至于所缺失的构件种类要比展示的种类多。举例说来，这里有扶梯，但是没有露台，壁炉作为一个加热和烹制的设备，能够将人们聚到一起，也被展示出来。但是相反，展览却从未提及建筑是怎样在房间内制冷的，类似的，关于屋顶的论述还提供了一些现代结构外壳中应用了特殊的中式覆盖物的案例。

这就好比库哈斯展示了面粉，鸡蛋，糖和黄油，然后告诉我们：这是一个蛋糕。所缺少的就是诗意地将这些原料转为结果的通道。

库哈斯的元素代表了一个冷门的尝试，即没有展示一位建筑师的实际工作，但却对建筑进行了仔细分析：设计空间的质量，要多亏这些以及其他元素的集合。

另一方面，尽管没有对空间的质量进行直接的讨论，但是如果展览在最后体现了策展人在其介绍文本中所分享的情感，即他与每个建筑构件之间的私人故事，那将会是十分有趣的。

And, in keeping with the Art Biennale that preceded it, curated by Massimiliano Gioni, The Encyclopedic Palace, Koolhaas provides a documentarian approach, an encyclopedia, in fact, of case studies, of reflections on the body of architecture, of references recalling those of a nineteenth-century manual expounded in spatial form. And in this expounding we see the limits of the curatorial choice. The chosen elements are in such abundance that what is lacking actually emerges more than what is presented: There are escalators, but, for example, no patios; the fireplace is shown as an element for heating, cooking, bringing people together, but the exhibition never mentions, by contrast, how architecture has often tried to cool rooms. Similarly, it does not seem possible that the discourse on roofs offers only examples of some specific Chinese covers entailed in modern structural shells.

It is as if Koolhaas presents flour, eggs, sugar and butter and tells us: This is a cake. What is missing is the passage that poetically transforms the ingredients into a course.

The Koolhaas's Elements represent an icy attempt to dissect architecture without showing the real work of an architect: to design a quality of a space, thanks to the assembly of these, and many other, elements.

On the other hand, even barring a direct discussion of the quality of the space, it would be interesting if the exhibition at least addressed those emotions the curator shares in his introductory text, the result of a very personal story of his relationship with the elements of architecture.

In this sense, the work of David Rapp, *Elements*, appearing at the entrance to the main pavilion, is able to convey, with extreme refinement, the empathetic relationship architecture creates between people and space: Through a tight assembly of hundreds of film clips, we get lost in a world where the Fundamentals are actually items that return an emotion to those who live in the architecture, that is, to all of us.

Rapp provides a brief moment of rupture from the attempt to

>> Monditalia+吸收现代性_军械库
Monditalia + Absorbing Modernity _ Arsenale

Monditalia展馆银狮奖
销售古怪性。Milano二区和电视城市化政策/安德烈斯·雅克（政治革新办公室）
1970年贝卢斯科尼着力推动了Milano二区的发展，并广而告之其为"最好的城市"。这座距离米兰7.5公里远的居住区，是为社会中的富裕阶层能够逃离污染、犯罪、工人阶层以及不稳定的混乱因素而设计的。贝卢斯科尼声明："我不销售空间，我只看业绩"。电视使房间卧室被描绘为Mediaset传媒集团扩建的Cigni宫殿；在这里性感的名人们将平常的生活方式解读为消费，吸引观众。Milano二区成为了试验品，在这里，二战后的国家电视台的城市化是跨媒体环境下全球面临的问题。

Silver Lion for the Monditalia section
Sales Oddity. Milano 2 and the Politics of TV Urbanism / Andrés Jaque (Office for Political Innovation)
In 1970 Berlusconi promoted Milano 2, a residential city 7.5 kilometers from Milan, advertised as "the city of the number one" and designed to exile affluent sectors of society, from pollution, crime, workers and uncertain market promiscuity. Berlusconi stated: *"I do not sell space, I sell sales"*. TV rendered living rooms as extensions of Mediaset's Palazzo dei Cigni; in which sexy celebrities translated ordinary life into consumption patterns, desirable to profiled audiences. Milano 2 has been the test-tube in which post-WWII national-TV-urbanisms were globally confronted by a transmedia environment.

照片提供：©Giorgio Zucchiatti

Monditalia展馆的特别提名奖
激进教学法：作用–反作用–相互作用/Beatriz Colomina, Britt Eversol, Ignacio G. Galán, Evangelos Kotsioris, Anna-Maria Meister, Federica Vannucchi, Amunátegui Valdés Architects, Smog.tv
该项目开发了一系列的教学实验，在20世纪下半叶塑造建筑学的理论和实践中扮演了至关重要的角色。作为对规范思维的一项挑战，这些实验质疑、重新定义并重塑战后时期的建筑领域。那些年里时时进出这个国家的流动的人口、书籍和观念使意大利的建筑、设计、教学法和激进性这些概念变得复杂化。展览描绘了一种新的空间地图网络，展现了规律再形成、教学方法和政治摩擦与交流间的空间关系。关注点在于空间的作用力、反作用力以及相互作用。

Special Mention for the Monditalia section
Radical Pedagogies: Action-Reaction-Interaction / Beatriz Colomina, Britt Eversole, Ignacio G. Galán, Evangelos Kotsioris, Anna-Maria Meister, Federica Vannucchi, Amunátegui Valdés Architects, Smog.tv
The project explores a series of pedagogical experiments that played a crucial role in shaping architectural discourse and practice in the second half of the twentieth century. As a challenge to normative thinking, they questioned, redefined, and reshaped the postwar field of architecture. The notions of "Italian" architecture, design, pedagogy and radicality were complicated during those years by the constant circulation of people, books and concepts out of and into the country. The exhibition draws a new kind of cartographic network of the spaces of disciplinary reformulation, pedagogical and political friction and exchange. The focus is on spaces of Action, Reaction, and Interaction.

从这个意义上来讲，戴维·拉普的作品《元素》出现在了通往主展馆的入口处，通过极其的精致性来传达建筑在人与空间之间创造的惺惺相惜的关系：穿过上百个电影胶片紧密结合形成的大门，人们迷失了，在迷失的世界里，基本法则实际上是能够将情感归还于那些生活在建筑中的人们，也就是我们所有人的条目。

拉普提供了一个简短的时间断层，尝试将一项只能是"主观"的规则变得"客观"，这也是因为自从建筑学（和艺术，如吉奥·庞蒂所强调的）不再取得进步时，我们不能像库哈斯曾努力尝试的那样，再将建筑学的表现方式限定在技术体系的进化中。

因此，建筑元素缺乏建筑学中最出类拔萃的元素：空间。从这层意思上来讲，它与理查德·柏戴特策划的第十届双年展很相似，被设计成一次大量关于城市研究的积累，与现代日本知名女建筑师妹岛和世策展的2010年展览的结果相距甚远，那届展览是一个对于建成空间的连续不断的研究。

相比之下，此次展览几乎是带着强有力的说教意义的：在每间展馆附近，我们都能目睹到其他一些还在进行中的项目，内部黑暗，到处散落着一些不用的材料。在揭示了建筑的技术元素之后，库哈斯看起来似乎想要通过向观众展示展览会的后台情景，来强调威尼斯双年展背后的辛苦工作。基于这种形式使人不禁开始怀疑，策展人真正的目标观众不是专家而是业余人士，展览通过将建筑分解成许多小部件而变得更加容易被消化，使观展者能够在这里开始熟悉和了解建筑学。

make "objective" a discipline that can only be "subjective", also because, since there is no progress in architecture (and art, as underlined by Gio Ponti), one cannot tether the narration of architecture only to the evolution of technology systems, as Koolhaas has attempted to do.
Elements of Architecture, therefore, lacks the element par excellence of architecture: space. In this sense it is very similar to the X Biennale, curated by Richard Burdett, which was configured as an accumulation of studies and research in the city, far away from the results of Kazuyo Sejima's 2010 show, which was a continuous succession of investigations of built space.
Here, by contrast, the exhibition is didactic, almost forcibly so: Around each exhibition room, we observe others, dark, populated with the unused materials of a continuous work in progress. It is as if Koolhaas, after revealing the technical elements of architecture, wanted to show the backstage of the exhibition itself, to emphasize the "dirty work" behind the Venetian show.
It is based on such gestures that one begins to suspect that the curator's real target audience is not professionals but non-experts, who can here begin to become familiar with architecture, made "digestible" through its decomposition into many smaller pieces.

Barbarians in the Arsenale

Quite different is the approach to the second stratigraphic extract, Monditalia, viewed along the Corderie of the Arsenale. Here Koolhaas's curatorship becomes a kind of tutelage, with much of the work delegated to OMA's Ippolito Pestellini Laparelli.
It is here possible to envision a different way of interpreting the research, a difference almost generational: While in Elements Koolhas develops an academic research, culturally linked to a typical mode of the twentieth century, Monditalia seems to present the strengths and weaknesses of contemporary cultural disclosure.
The writer Baricco, in his essay of *Mutation of Culture*, 2006, ironically defines Barbarians who are increasingly bypassing the dic-

乡村小礼拜/马蒂尔德·卡萨尼

锡克教在意大利波河平原的乡村地区建造了很多庙宇，招待每年来参加大型收获节日——丰收节的人们，为了这个节日，上千名锡克教教徒会聚集在这里。在那几天里，在这个村子里意大利的宗教景观一下子变得明显起来。记得在这20年里，意大利已经从一个大多数信奉天主教的国家变成了以一个复杂的且空前的以多样性宗教形式为特点的国度。丰收节质疑着城市的变化和以宗教为目的的空间利用如何能比意大利任何城市的规划政策、工具和条例的变化都要多。

Countryside Worship / Matilde Cassani

Every year the many temples Sikh built across the Italian agricultural farmlands of the Pianura Padana play host to a huge harvest festival, the Vaisakhi, where thousands of Sikhs congregate for the event. In these days, the Italian religious landscape suddenly becomes evident, within its countryside, remembering that in twenty years, from a country with a Catholic majority Italy to become characterized by a complex, unprecedented pattern of religious diversity. The Festival questions how urban change and the use of space for religious purposes happen much more swiftly than any change in Italian urban planning policies, tools and regulations.

图片提供：©Chiara Quinzii

Monditalia展馆的特别提名奖_意大利边界/Folder

由于在现代主权国家的定义中起到了根本性作用，意大利的高山、河谷和岬角常常被特定的控制系统改变和移植。在2008至2009年间，尽管一次毫无预兆的改革迫使意大利政府不得不与毗邻的欧洲国家重新协定边界。由于高山冰川的融化，长长的、绵延的、决定边界线的分水岭始终在移动。“移动的边界”这个新概念被引入国家立法中，反驳了关于意大利边界精确定位的任何必然性。

Special Mention for the Monditalia section _ Italian Limes / Folder

Italy's peaks, valleys, and promontories have always been altered and colonised by peculiar systems of control that played a fundamental role in the definition of the modern sovereign state. Between 2008 and 2009, though, an unexpected transformation forced the Italian government to negotiate a new definition of its frontiers with adjacent European countries. Due to shrinking Alpine glaciers, the watershed – which determines large stretches of the border line – has shifted consistently. A new concept of movable border has thus been introduced into national legislation, refuting any certainty about the precise position of Italian borders.

军械库内的野蛮人

Monditalia展馆的与众不同之处是其对第二地层的勘探，库哈斯的职务由管理者变成监督人，他将许多工作都委托给了OMA的Ippolito Pestellini Laparelli来完成。

在这里，可以设想以一种几乎跨越了一个时代的差异的方式来解释研究结果：在the Elements展馆，库哈斯开发了一种学术研究，从文化的角度来看与20世纪的典型风格联系到了一起，而Monditalia展馆似乎是代表了现代文化所暴露出来的优势与不足。

作家巴里科在其2006年发表的文章《文化的突变》中，讽刺地描述了野蛮人越来越多地绕过浪漫文化的规定，自19世纪持续至今，新型野蛮人将文化变成了信息的积累，只做表面功夫而不做深度探寻，更关注相同信息（即使是看上去价值不高的信息）之间的关联，而不屑于对同一话题进行长久而耐心的研究。

Monditalia展馆中的情形非常相似：数不尽的项目堆积、各种不同的媒体、几近混乱的参考文献。如同汤博乐网站，展览并不在其零散的信息碎片间寻求绝对的一致性：总体叙事线条薄弱、清晰、私人化，每个展出的项目都潜藏着有趣的信息却无从表达，这正是因为信息总量过大而难以充分辨别。

在众多的工程中，这六项也许是最有趣、最吸引人的：

——民主剧院，荷兰XML研究城市化建筑师事务所设计，研究整个历史时期政治结构下的建筑形式。

——99多米诺，Space Caviar事务所设计，以99座意大利建筑为例

tates of romantic culture, as it has persisted from the nineteenth century to the present day: The new barbarians make culture an accumulation of information, to be navigated on the surface and not in depth, focusing on connections between the same information (even seemingly feeble ones), rather than on long and patient study of a single topic.

And something similar is what happens in Monditalia: an accumulation of countless projects, a variety of media, a superfluity of references that verges on chaos. Like a Tumblr site, the exhibit does not seek absolute consistency between its pieces of information: The binding narrative thread is quite thin and clearly partial, personal – each presented project has interesting potentialities that remain unexpressed due to an amount of data so huge as to be impossible to fully discern.

Among these projects some of the most interesting and absorbing are perhaps these six:

– Theaters of Democracy, by XML, which investigates the architectural forms of political assemblies throughout history;

– 99 Domino, by Space Caviar, which displays a kind of declination of the projects of Le Corbusier through 99 examples of Italian building;

– Countryside Worship, by Matilde Cassani, who through two large lenticular images is able to animate the urban void portrait in the photo;

– Radical Pedagogies, by Beatriz Colomina with PhD students of Princeton University, which shows, with great emphasis, the actual pedagogical and didactic approach of the Biennale;

– Z! Zingonia, mon amour, by Argot ou La Maison Mobile + Marco Biraghi, which gives the impression of breaking through the dark spaces of the Arsenale with three large skylight-lamps;

– Italian Limes, by Folder, which studies the concept of the "moving boundary" through the glaciers of the Alps.

Z! 欣歌尼亚, 我的爱/Argot ou La Maison Mobile, Marco Biraghi

欣歌尼亚是现代建筑乌托邦的一个固有形象，它是介于阿德里亚诺·奥利弗蒂的“伊夫雷亚小镇”和Archizoom的“无休止的城市”之间的一个典范性时刻。忽略这座建筑目前的状态，它曾是由Renzo Zingone设计并由建筑师Franco Negri建造的，混合了激进的城市设计的种子和幻想的建筑风格。关于该项目的出版物由Zingone Iniziative Fondiarie授权出版，它是一个装满了20世纪60年代的美好愿望的时间胶囊，一本有关意大利的企业家在其鼎盛时期的欲望、野心和梦想的日记。双年展展出的欣歌尼亚是为了防止其文化根基遭到忽略；被遗忘是每个现存的乌托邦所面临的真正风险。欣歌尼亚现在正在经历一个更新换代的过程，这也是所有近代和当代建筑所面临的共同命运：转变还是建造。伦巴第地区在其建成的50年后，为了与Verdellino直辖市及Cisarano直辖市保持一致，也插入了一项改进性措施，允许城市的替代部分被彻底分化。Z! 欣歌尼亚展馆表现了建筑的脆弱之处，天性沉重且刚性的结构由社会、经济、政治和法律这四个不断变化的柱子支撑着。

Z! Zingonia, Mon Amour/Argot ou La Maison Mobile, Marco Biraghi

Zingonia is a fixed image in the blur of modern architectural utopias, an exemplary moment between Adriano Olivetti's Ivrea and Archizoom's No-Stop City. Despite its current state, the city designed by Renzo Zingone with architect Franco Negri incorporates the seeds of radical urban design and visionary architecture. The original publication of the project, commissioned by Zingone Iniziative Fondiarie, is a time capsule full of good intentions straight from the 1960's, a diary of the desires, ambitions and dreams of Italian entrepreneurship at its best. Zingonia is on exhibition at the Biennale to prevent the neglect and decay of its cultural roots: being forgotten is the real danger for every living utopia. Zingonia is currently undergoing an updating process, which is the common fate of all modern and contemporary architecture: conversion vs. construction. Fifty years after its foundation, the Lombardy Region, in accordance with the Municipalities of Verdellino and Cisarano, is intervening with an innovative approach that will allow the replacement of parts of the city now totally degraded. The Z! pavilion embodies the fragility of architecture, by nature a heavy and rigid element that rests on four constantly shifting pillars: society, economics, politics and law.

民主剧院/XML

在“民主剧院”项目中，XML建筑事务所研究了半圆形的古希腊露天剧院这座为政治集会而建造的建筑类型。研究确定“观众即为公民”这个概念是古希腊露天剧场与今天的国会间的一项共同特性。在Corderie展馆中放置了一个双面物体。一面贴着的壁纸代表对于古希腊露天剧院的研究和其在古希腊城邦的直接民主制中所扮演的角色。而在其他壁纸中，XML通过系谱来追溯半圆的轨迹，进而展示今天的民主性——而希腊剧院作为一个强有力的民主标志一直坚挺着。而实际上民主却发生了很大的变化。

在墙体的另一面包含了一系列的窥视孔，可以窥见如今的议会大厅；体验如今的政策vs.希腊剧院部分模型（部分位于墙体的另一侧）的感觉。

然而，展览的一部分已被集体所认知（作为巨型墙纸面前的组合矗立着），且人们在其他部分观看材料的方式也较为私人化。

Theaters of Democracy/XML

In the project "Theatres of democracy" architecture office XML researches the typology of the semi-circular Greek theatre as an architectural device for political congregation. The research identifies the notion of "spectatorship as citizenship" as a common denominator between the ancient Greek theatre and today's parliaments.

The installation in the Corderie consists of an object with two sides. On one side, a wallpaper presents research into the Greek theatre and its role in the direct democracy of the Greek polis. In a second part of this wallpaper XML traces through a genealogy how the semi-circle has travelled to present today's democracies - whereas the Greek theatre as a powerful symbol of democracy persisted, the reality of democracy has changed significantly.

The other side of the wall consist of a series of peepholes into today's halls of parliaments: experiencing the feeling of exclusion of today's politics vs. the inclusive model of the Greek theatre with part of the installation on the other side of the wall. Whereas one part of the exhibition is perceived collectively (standing as a group in front of a giant wallpaper), the way that people can look at the material in the other part is individual.

演示了建筑大师勒·柯布西耶作品的衰落。

——乡村小礼拜，马蒂尔德·卡萨尼设计，通过两幅巨大的透镜影像能够使照片中枯燥的城市肖像活泼生动起来。

——激进教学法，由建筑历史学家比阿特丽斯·哥伦米娜与普林斯顿大学的博士生合作完成，着重展示了威尼斯双年展的教学和说教方法。

——Z!欣歌尼亚，我的爱，Argot ou La Maison Mobile和Marco Biraghi设计，通过三个巨大的天窗–照明设备给观展者留下了突破军械库黑色区域的印象。

——意大利边界，Folder建筑事务所设计，通过阿尔卑斯山脉的冰川来研究“移动的边界”概念。

相比主展馆，在Monditalia展馆中，整体的展览设计展现了其内容，事实上，一些诗情画意的瞬间为Corderie的沿线空间赋予了某种意义。由于承办由艺术家们来演绎的舞蹈、音乐、戏剧和电影的双年展，这些瞬间被植入其中。在这里又有大量的信息流（除了个别项目中有很长历史的电影游行在意大利进行），停于此处为人们提供休息空间，游客可以享受这项运动，以及步道环绕的舞台上的艺术家说出的台词，在这里人们可以从多个视角观看表演。仿佛空间真地活跃了起来，展示其真正的本质——不仅仅是研究，更是一种观展者可以通过皮肤触碰到的一种感觉，回溯到那个问题上，一次建筑展览应该是：对空间进行的一次实验，以及对这个空间赋予资格的过程。

In Monditalia, by contrast with the Main Pavilion, the general exhibition design takes over the content and, in fact, has some poetic moments which give some sense to the space along the Corderie: They are inserts constructed to host performances by artists associated with the Dance, Music, Theatre and Cinema Biennales. Here the flows of information, again superabundant (in addition to individual projects there is a long parade of historical films set in Italy), stops to accommodate breaks and allow visitors to enjoy the movement and words of artists who perform on small stages surrounded by walkways, on which it is possible to find multiple views of the performances.

It is as if the space finally comes alive, showing itself in its real essence – not only studies and research, then, but a sensation one can feel on the skin and that goes back to what an exhibition of architecture should be: an experiment on the void, a qualification of that same void.

History Manuals

Absorbing Modernity: 1914-2014, the theme that Rem Koolhaas has given to the national pavilions, has generally led, to a historicist approach, still based on the collection of documents and testimonies of a recent past.

Many pavilions have thus tried to retrace the steps of modernity, taking a backward glance, becoming chests of three-dimensional local history courses: Due to the strong curatorship imposed by the Dutch architect, only a few pavilions exhibit spatial research or attempt to show experiments with it; many pavilions have decided to offer documentary research, escaping discussion of the future as if in fear of being unable to justify their curatorial choices. What emerges is, in fact, a portrait of a world frightened of prefiguring the future, almost as if waiting for responses to an indefinite crisis that threatens to reconstruct certainties that nevertheless seem to offer more dead weight than useful reflection in imagin-

金狮奖_朝鲜半岛鸟瞰图

经历第二次世界大战的瞬时巨变之后，朝鲜半岛被一分为二。在冷战两极分化的全球和国家体制内，一个维持国家统一实体长达千年的社会和文化逐渐变得迥异，但却又不可避免地在经济、政治和意识形态体系方面相互联系。在韩国馆内，朝鲜和韩国的建筑风格作为一个统一的媒介呈现在人们面前——一个以一种全新的方式产生交替叙述的机制，它既能够使人感知寻常日子又能够感知那些值得纪念的日子。

韩国馆的设计受到了由建筑师转作诗人的李箱（1910年—1937年）创作的《乌瞰图》这首诗的启发。与单一和普遍的鸟瞰视角相比，《乌瞰图》指出对分裂朝鲜建筑风格以及对建筑理念本身的整体把握的不可能性。好像不平整的地球的未知土地一样，由建筑师、城市规划专家、诗人和作家、艺术家、摄影师、电影制作人、策展人和收藏家创作的不同作品形成一套多样的研究项目、入口节点和视角。他们呼吁人们关注规划的以及非正式的、单一以及集合的、史诗般以及日常的城市和建筑现象。

韩国馆既将朝鲜半岛展示为一种现象，也把它作为一个媒介，既是过去100年以来全球动荡轨迹的原型，同时也是一个特例。

Golden Lion_ Crow's Eye View: The Korean Peninsula

In the immediate aftermath of World War II, the Korean Peninsula was divided in two. Within the polarizing global and state systems of the Cold War, a society and culture that had maintained a unified state entity for more than a millennium evolved radically divergent yet irrevocably interconnected economic, political, and ideological systems. In the Korean Pavilion, the architecture of PRK and Korea is presented as an agent – a mechanism for generating alternative narratives that are capable of perceiving both the everyday and the monumental in new ways.

The Korean Pavilion is inspired by "Crow's Eye View," a poem by the Korean architect-turned-poet Yi Sang (1910 -1937). In contrast to the singular and universalizing perspective bird's eye view, the crow's eye view points to the impossibility of a cohesive grasp of not only the architecture of a divided Korea but the idea of architecture itself. Like uncharted patches of an irregular globe, a diverse range of work produced by architects, urbanists, poets and writers, artists, photographers and film-makers, curators and collectors forms a multiple set of research programmes, entry nodes, and points of view. They call attention to the urban and architectural phenomena of the planned and the informal, individual and collective, the heroic and the everyday.

The Korean Pavilion reveals the Korean Peninsula as both symptom and agent, both archetype and anomaly of the tumultuous global trajectory of the past 100 years.

照片提供：©Italo Rondinella

患者의容態에관한문제。

1234567890·
123456789·0
12345678·90
1234567·890
123456·7890
12345·67890
1234·567890
123·4567890
12·34567890
1·234567890
·1234567890

診斷 0·1

26·10·1931

以上 責任醫師 李箱

李箱，1934年创作的《4号诗，乌瞰图》 2014年由Sulki和Min排版

Yi Sang, "Crow's Eye View, Poem No. 4," 1934; typeset by Sulki and Min, 2014

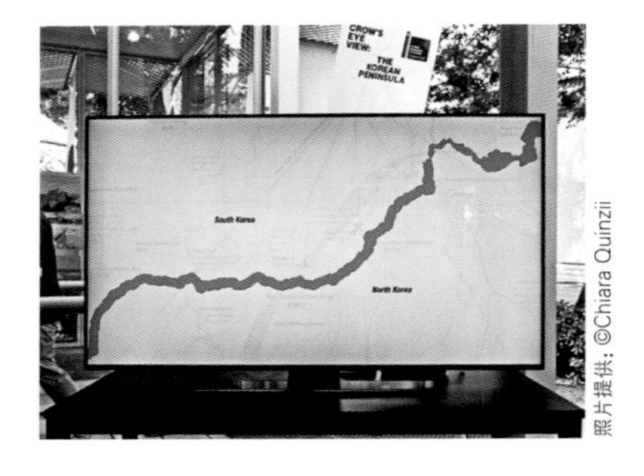

照片提供：©Chiara Quinzii

历史指南

吸收现代性：1914—2014，雷姆·库哈斯赋予国家场馆的主题已引导出立足于通过收集近期历史文献和证据的方式来进行广泛使用的历史主义方法。

许多场馆已因此尝试追溯现代性的步调，回顾过去，成为当地真实历史进程的三维载体；因受到荷兰建筑师作为强势策展人施加的主题的影响，仅有几个场馆展示了空间研究或尝试展示空间研究实验；许多场馆都已决定提供文献研究，它们似乎在担心不能够证明自己的策展选择，因而对建筑的未来避之不谈。事实上，这体现出来的是人们对预想的未来世界的畏惧，就像是在等待对威胁重建某栋建筑的不确定危机做出回应，然而这种回应在憧憬未来方面似乎带来更多的是累赘而非有益的影响。正因为如此，许多场馆开辟了一条与OMA方法非常相似的道路，那就是谈论其他话题，使多门学科参与其中以支持单调乏味的建筑，在不添加任何数据、信息以及建筑空间的外部参考的情况下，建筑既不能证明自己也不能表现其自身的价值。

于是，金狮奖颁给了由曹敏硕担任策展人的韩国馆，该馆在讲诉突出的政治事件方面优于其他场馆，在该馆中，建筑又是一个讨论其他话题的理由。正如策展人所说，它不仅只涵盖建筑：在39名工作人员当中，

ing the future. It is for this reason, then, that many pavilions chart a course very similar to the OMA approach, which is to talk about something else, to involve a multiplicity of disciplines as though to support often plodding architecture that can neither justify itself nor express its value without the addition of data, information, and references external to the architectural space.

Accordingly, the Golden Lion has been awarded to the Korean Pavilion, curated by MinSuk Cho, which has managed better than others to tell a story that is eminently political, in which the architecture is again merely an excuse to talk about something else. As the curator remarks, it's not only about architecture: among 39 of them, only 19 are architects, urbanists, architecture theorists and people from our display. The rest are all through different disciplines such as art and filmmaking. The exhibit's studies and research overlap the architecture on tangentially, as the jury confirms: It is a research-in-action, which expands the spatial and architectural narrative into a geopolitical reality.

In fact there is obvious ambiguity in the exhibition and in the question, unanswered, whether architecture has influenced civil society or vice versa. This ambiguity maintains the pavilion's balance by allowing visitors to become lost in images – monumental in the case of PRK and chaotic in the case of Korea – or in the his-

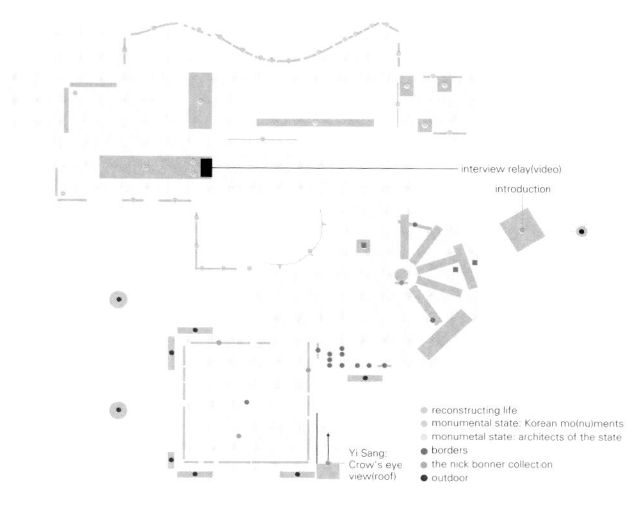

> 重建生活 Reconstructing Life

照片提供：courtesy of Peter Blum Gallery

1957年由克利斯·马克拍摄的跑步穿过金日成广场的女孩，"朝鲜人，无题 #16"
Chris Marker, Girl running through Kim Il-Sung Square, "Koreans, Untitled #16," 1957

照片提供：courtesy of Arts Council Korea

照片提供：courtesy of Arts Council Korea

> 不朽的国家 Monumental State

照片提供：©Osamu Murai

1984年金寿根在首尔奥林匹克运动场
Kim Swoo-Geun at the Seoul Olympic Stadium, 1984

> 边界 Borders

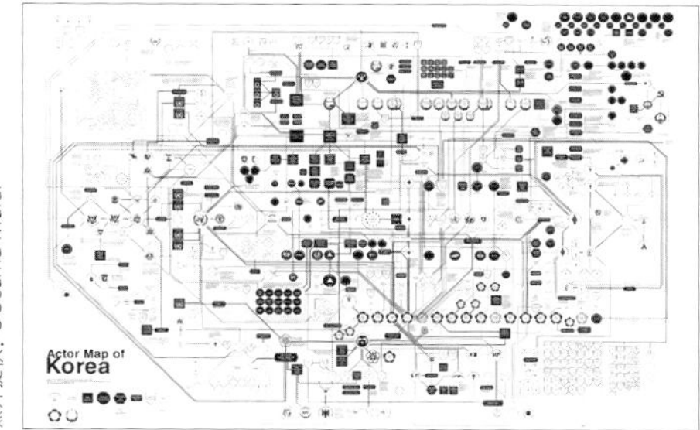

Suh YehRe于2014年创作的韩国参与者地图
Suh YehRe, Actor Map of Korea, 2014

> 理想的旅行 Utopian Tours

照片提供：courtesy of Nick Bonner

学院建筑，匿名建筑师建造，纸上建造的有机玻璃模型，2011年
Academy of Architecture by anonymous architect 2011; acrylic on paper

只有19人是建筑从业人员、城市规划专家、建筑理论家以及展览团队人员。其他人员全部来自不同的行业，如艺术和制片行业。展览品的研究和调查将与建筑学的重叠之处一带而过。正如评委会所评论的：它是一次研究行动，将空间和建筑叙述扩展至地缘政治的现实当中。

事实上，展览中存在着明显的不明确之处，而且存在着尚未得到解答的问题，到底是建筑影响了公民社会，还是相反的情形。而这种不明确通过使参观者以沉浸于图海的方式来维持场馆的平衡——对于朝鲜是一些具有纪念性的图片，而对于韩国则是一些杂乱无章的图片——或是使人们迷恋于在两国边境的历史，再一次在未被要求的情况下研究设计空间的质量。

总之，我们观察到韩国的策展理念总体上可以被视作是两个被迫平行世界的发展，以菲利普·K.狄克就不一样的未来的讨论所反映出来的便是：这些问题似乎是通过哲学而非建筑学推测而得到解决。

然而，最有趣的是将朝鲜半岛作为一个统一而非分立地区来观察的可能性，将建筑学描述为社会和政治局势的一种衰败。该馆没有将建筑学拆分至其基本的要素，在这里参观者可以看见各个行业的分类，这些行业影响空间并同时受到空间的影响。

然而，在花园的设计中，参观者可以注意到一种不同的方法，设计者也

tory of the boundary between the two nations, once again without ever being called upon to analyze the quality of the designed space.

We observe, in short, the idea that Korea, in its entirety, can be seen as the development of two forcedly parallel universes, as born from a reflection of Philip K. Dick in discussing an alternative future: these matters seem be addressed via philosophical, rather than architectural, speculation.

What is of great interest, then, is the possibility of observing the Korean peninsula as a place of unity rather than division, describing the architecture as a declination of social and political situations. Instead of breaking the architecture into its fundamental elements, here one can see a breakdown of disciplines that simultaneously affect and are affected by the space.

In the Gardens can be noted, however, a different approach, one that perhaps declines Koolhaas's demands in favor of a more proactive architectural vision:

– the Belgian pavilion, which constructs a space of exciting lightness, with thin white metal frames that accompany the exhibition and in which the absorbed modernity is conveyed, ironically, by the presence of some household refrigerators along the pavilion wall;

– the Bahrain pavilion, where the history and the documents that

照片提供：courtesy of the Chilean Pavilion (©Gonzalo Puga)

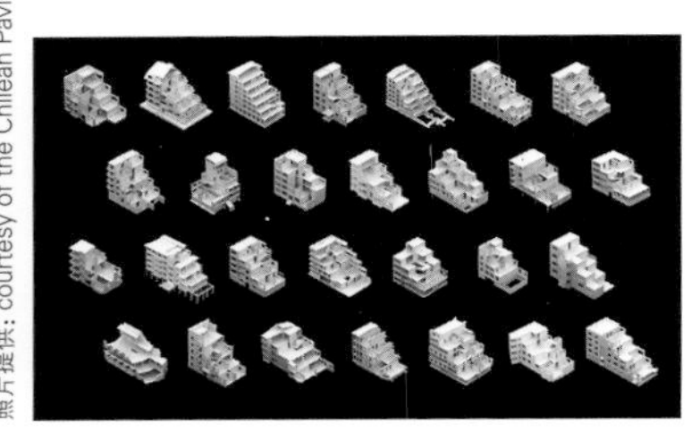

照片提供：courtesy of the Chilean Pavilion

银狮奖_巨石的争议
在第14届威尼斯建筑双年展上，智利馆的一块大型混凝土板独立于该馆的垂直位置。它是KPD工厂（生产预制装配式住宅建筑构件的工厂）生产的首批产品之一，由前苏联于1972年捐赠给倡导社会主义道路的智利总统萨尔瓦多·阿连德。这块板从那以来便成为诸多政治和思想争论的作用剂，尤其是在阿连德在塑性混凝土板上签名之后，如果他的这项举动后来没有被奥古斯托·皮诺切特的专制行为（在殖民时期风格的吊灯配件中间为该板增加圣母和圣婴的描绘）掩盖的话，估计它会一直是争论的作用剂。这块混凝土板代表了现代建筑编史中一个相对临界的传统，尽管在20世纪下半叶中全球建造了超过1亿7000万的大型混凝土板公寓。该板由此成为第14届威尼斯建筑双年展提出的吸收现代性主题的基本标志。
“巨石的争议”是由佩德罗·阿朗索和雨果·帕尔马罗拉共同创作的作品，该作品是基于一项记载KPD工厂于智利建造的153栋住宅的研究项目，以及1931年至1981年之间全球范围内开发和散布的28块大型混凝土板系统的技术性、类型化和概念性重建的。

Silver Lion _ Monolith Controversies
Isolated in upright position, a large-concrete panel stands in the center of the Chilean Pavilion at the 14th International Architecture Exhibition of la Biennale di Venezia. This was one of the first ever produced by the Chilean KPD Plant – an industry to produce prefabricated housing donated in 1972 by the Soviet Union to the Chilean road to socialism led by president Salvador Allende. This panel has since been the agent of several political and ideological controversies, especially after Allende himself signed it up in the wet concrete, if only for his gesture to be later covered up by Augusto Pinochet's dictatorship, adding the panel the representations of the Virgin and Child between two colonial style lamp fixtures. This piece of concrete represents a relatively marginal tradition in the historiographies of modern architecture despite the fact that more than 170 million large-concrete panel apartments were built worldwide during the second half of the twentieth century. Thus the panel becomes fundamental symbol to the absorption of modernity proposed as the concept for the 14th International Architecture Exhibition.
Monolith Controversies is a work by Pedro Alonso and Hugo Palmarola based in a research project documenting the 153 housing blocks built in Chile by the KPD Plant, as well as the technical, typological and conceptual reconstruction of twenty-eight large-concrete panel systems developed and disseminated worldwide between 1931 and 1981.

许拒绝应承库哈斯的要求，而是采取一种更加具有前瞻性的建筑视景：

——比利时馆建造了一处极其明亮的空间，展览搭配白色薄金属框架，并且在展览中通过沿场馆墙壁设置一些家用冰箱来批判性地传达吸收现代性的主题；

——在巴林场馆内，历史以及验证历史的文献得到了立体的表现，形成一个巨大的圆柱形图书馆，似乎能以一系列的文字和声音来吸引参观者；

——在摩洛哥馆，参观者可以根据楼层来领会不同方式的步行体验：沙地在软硬过渡的过程中形成了一时的轻微失和，且占据房间整体表面的投影帮助提升了视域；

——泰国馆营造了一处神奇并且神秘的氛围；

——塞尔维亚馆，在室外黑暗和室内亮度之间形成暗示性的对话；

——德国馆在场馆自身空间中重建了一栋私人别墅，在私人与公共场所之间营造一种几乎超现实的效果。

最后一个有趣的场馆位于普里奇欧尼宫临近圣马可广场的位置：台湾馆的建造似乎基于20世纪70年代意大利激进分子开展的实验，该馆设计有嵌入历史建筑的彩色微构架，而小型创意空间向人们诉说典型家庭拥有的一个个独立的故事。

在这里，建筑仿佛不是通过库哈斯的要素，而是通过定义环境的行为来进行分解，以设法将空间研究的兴趣重新带回给那些真正欣赏建筑质量的人们。

testify to it acquire three-dimensionality, becoming a giant cylindrical library that seems to attract visitors with a flurry of words and sounds;
– the Morocco pavilion, where one can appreciate the experience of different ways of walking depending on the floor: Sand creates a moment of slight estrangement in the transition from hard to soft, and the unpublished upward vision facilitated by a projection on the total surface of the room;
– the Thailand pavilion, which creates a magical and mysterious atmosphere;
– the Serbian pavilion, with its suggestive dialogue between outer darkness and inner luminosity;
– the German pavilion, which reconstructs a private villa in the spaces of the pavilion itself, creating an almost surreal effect between the private and public.

The last interesting pavilion sits near the Piazza San Marco, in the Palazzo delle Prigioni: The Taiwan Pavilion seems to build upon the experiments of the Italian radicals of the 1970s with its colored micro-architectures inserted in the historic building, and small exciting spaces that tell individual stories of typical household uses. It is as if the architecture here was decomposed not via the elements of Koolhaas, but through the actions that define environments, managing to bring the interest of the spatial research back to the very people who actually enjoy the quality of the architecture. Diego Terna, Photographs by courtesy of la Biennale di Venezia(Elements of Architecture & Monditalia: ©Francesco Gaili / Absorbing Modernity: ©Andrea Avezzù), Except as noted

>> 法国馆 The French Pavilion

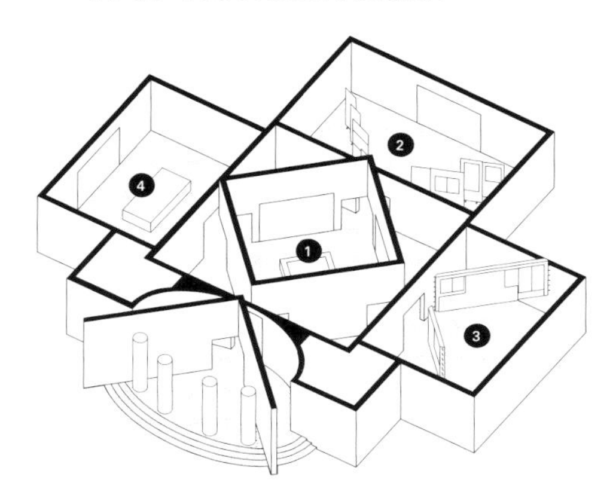

1 雅克·塔蒂和Arpel别墅：是欲望的客体还是可笑的机器?
2 让·普鲁韦：是建构的想象力还是乌托邦?
3 大量预制组件：是规模经济还是单一经济?
4 大型住宅区：是拯救的异托邦还是隐居之所?

1. Jacques Tati and the Villa Arpel: object of desire or machine of ridicule?
2. Jean Prouvé: constructive imagination or utopia?
3. Heavy prefabrication: economies of scale or monotony?
4. The large housing estate: heterotopia of salvation, or place of reclusion ?

特别提名奖_现代性，是承诺还是威胁?

1914年后，法国在自身的建设过程中，建筑师和工程师大多是为了满足社会各阶层的需求而做出特殊贡献，他们似乎从未真正地接受"现代性"。虽然在其他国家的舞台中，建筑现代性和社会改革的相遇带来了巨大的希望，例如为残疾人提供通往住宅以及公共服务设施的入口。然而这种相遇形成了自己的形式。尽管倾向于国际化和英国、德国或美国模型带来的影响力，自1914年以来，法国就以确保社会进程的重要公共政策的循环影响为其特点，并且以1928年西格弗里德·吉提翁提出的"建筑气质"作为法国国家的"定义性特征"。这一结合促进构成了1928年Loucheur法将国家资金注入住宅建设的项目、3个重建项目以及1950年之后的大型住房开发建设政策。第二次世界大战没有使世界处于停滞状态，反而加速了现代化的进程。松散的国家边境使英国、德国和美国的建筑模型、形式和技术在这一时期大量涌入，同时法国在阿尔及利亚和摩洛哥这两处的殖民实验室也开展了自己的实验，随即很快将其转移至大都市中。

Special Mention_ Modernity, Promise or Menace?

Since 1914, France has not so much absorbed modernity as has shaped it, through its distinctive contributions by architects and engineers responding to the expectations of different components of society. As on many other national scenes, the encounter between architectural modernity and social reform raised great hopes, such as providing the disadvantaged with access to proper housing and to collective support structures. Yet this encounter took its own forms. Despite the tendencies toward internationalization and the power of the British, German or American models, the French scene since 1914 has been characterized by the recurring influence of important public policies aimed at ensuring social progress, and by the "constructional temperament" that Sigfried Giedion had already set out in 1928 as a "defining feature" of the French nation. This fertile conjunction established the programs for the Loucheur law of 1928 that funneled state funding into housing, the programs for the three reconstructions, and the building policies of large housing developments after 1950. Far from being a time of stasis, the World War II accelerated the process of modernization. National borders revealed themselves to be porous, with models, forms and techniques continuing to arrive from Britain, German and the United States throughout this period, while France conducted its own experiments in the colonial laboratories of Algeria and Morocco, which would soon be transposed to the metropolis.

>> 加拿大馆 The Canadian Pavilion

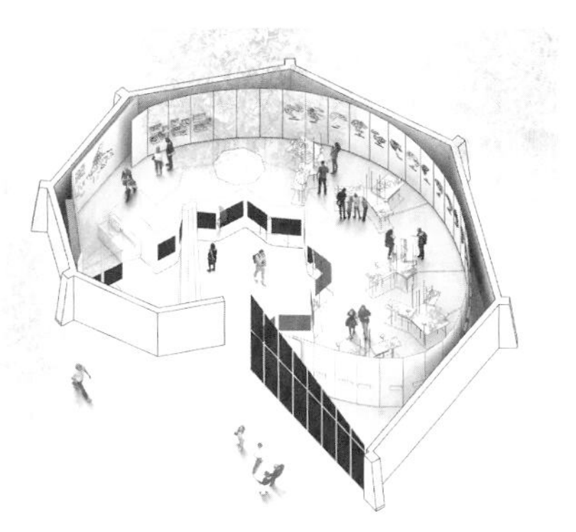

照片提供：courtesy of the Canadian Pavilion

特别提名奖_适应北极：努那伍特的15年

适应北极：《努那伍特的15年》调查了近年来建筑的过去、当前的城市化现状以及对不远的将来努那伍特适应性建筑的映射。努那伍特，意为"我们的土地"，是加拿大最新、最大以及最北边的领土。根据1993年经过努力争取而达成的领土声明协议，该地于1999年4月1日与西北领地分离开来。如今，大约有33 000位居民生活在这片占地200万平方公里的土地上，他们遍布于25个社区中，使努那伍特成为世界上人口最稀少的地区之一。这些社区地处森林线之上，并且不与公路连接，人口数量从最小面积的哈姆雷特区的120人到努那伍特首府伊魁特的7000人。努那伍特的气候、地理和人口，以及更加广阔的加拿大北极地区，挑战着日趋普遍的现代性。

随着20世纪两极探险时代的来临，现代建筑以主权、原著民事务管理、或是贸易以及其他名义吞食着加拿大的广阔地区。然而，原住民因纽特人作为传统的半游牧民，已在加拿大的北极地区居住了千年之久。

适应北极，这一设计直接响应了第14届威尼斯建筑双年展的主题。现代性通常畏惧地区的特异性和"地方性"这一前提。努那伍特似乎仍抗拒现代主义普遍化的趋势。这次独特的展览力图展露建筑的抵制行为，并让人们认识一个不为人知的现代化的加拿大北极地区。

Special Mention_ Arctic Adaptations: Nunavut at 15

Arctic Adaptations: Nunavut at 15 surveys a recent architectural past, a current urbanizing present, and a projective near future of adaptive architecture in Nunavut. Nunavut, which means "our land", is Canada's newest, largest, and most northerly territory. It separated from the Northwest Territories on April 1, 1999 following a hard-fought land claims agreement established in 1993. Today, there are almost 33,000 people living in 25 communities across two million square kilometres, making Nunavut one of the least densely populated regions in the world. These communities, located above the tree line and with no roads connecting them, range in population from 120 in the smallest Hamlet to 7,000 in Nunavut's capital city of Iqaluit. The climate, geography, and people of Nunavut, as well as the wider Canadian Arctic, challenge the viability of a universalizing modernity.

Following the age of polar exploration in the 20th century, modern architecture encroached on this remote and vast region of Canada in the name of sovereignty, aboriginal affairs management, or trade, among others. However, the indigenous Inuit people have inhabited the Canadian Arctic for millennia as a traditionally semi-nomadic people.

Arctic Adaptations responds directly to the theme of the 14th Venice Architecture Exhibition. Modernity is often fearful of the specificities of place and the premise of "the local". Yet Nunavut seems to resist modernism's universalizing tendency. This unique exhibition seeks to reveal acts of architectural resistance and identify an unrecognized modern Canadian North.

>> 俄罗斯馆 The Russian Pavilion

特别提名奖_足够公平：俄罗斯的过去，我们的当下

1914年以来，俄罗斯已将自身建立为世界上最大且最激进的城市实验室。数十年的实验进程产生了几乎能够满足所有人口需求或社会野心的建筑解决方案。俄罗斯为了这些努力付出了沉重的代价，而它们的有效性往往低于所付出代价的成本。"足够公平"这一主题没有呈现出俄罗斯现代化建设的线性故事，而是运用建筑历史以满足当代的需求。该展览从过去的一个世纪中汲取城市的理念，有些是值得赞扬的，有些是不为人知的；有些看起来已经过时，而有些据说是失败的，并给予它们新的目的。

为使其效用实现最大化，每一个展出的项目都被精简至其概念性的本质。为阐明它们之间持续的相关性，这些理念进行了更新，并用于挑战当下世界各地的建筑师。在最简单的交流方面，该馆则通过国际贸易博览会的通用语言将俄罗斯的建筑的创新性表达出来。

该馆的场景本身就是对国家场馆策展主题的一种响应：此次展览通过应用可能是现代性最终表现形式的语言，提出了"吸收现代性"的问题。"足够公平"是一个理念的展示会。每一个展品都是现代化过程中的一个里程碑，并且为新的努力开辟道路。它们共同形成了一个城市发明的市场——俄罗斯制造，向世界开放。

Special Mention _ Fair Enough: Russia's Past, Our Present

Since 1914, Russia has established itself as the world's largest and most radical urban laboratory. Decades of experimentation have produced architectural solutions for almost any demographic need or social ambition. These efforts are undertaken at great cost, and their usefulness is often undersold. Rather than presenting a linear story of Russia's modernization, Fair Enough applies architectural history to meet contemporary needs. The exhibition takes urban ideas from the past century – some celebrated, some obscure; some seemingly outdated, some supposed failures – and gives them new purposes.

To maximize its utility, each exhibited project is stripped to its conceptual essence. To illustrate their continued relevance, the concepts are updated and applied to challenges now confronting architects around the world. For easiest exchange, Russia's architectural innovations are expressed through the universal language of the international trade fair.

The scene itself is a response to national pavilions' curatorial theme: the exhibition addresses the issue of "absorbing modernity" by adopting the language of what is perhaps the ultimate manifestation of modernity. Fair Enough is an expo of ideas. Each exhibit marks a milestone in modernization and clears a path for new efforts. Together, they form a marketplace of urban invention – made in Russia, open to the world.

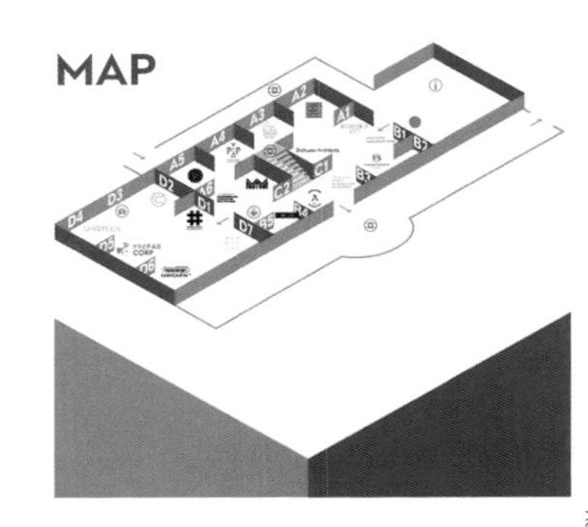

照片提供：©The Archive of A.M. Shchusev

休谢夫：Narkomzem大楼（莫斯科，1928年）
Shchusev: Narkomzem Building (Moscow, 1928)

A1展位：Estetika Ltd，A2展位：Lissitzky
Booth A1: Estetika Ltd, Booth A2: Lissitzky

B4展位：Archipelago Tours
Booth B4: Archipelago Tours

照片提供：©Peretz Partensky

Archipelago Tours：位于阿富汗马扎里仍处于运作状态的面包工厂
Archipelago Tours: still-functioning bread factory in Mazar, Afghanistan

>> 塞尔维亚馆 The Serbian Pavilion

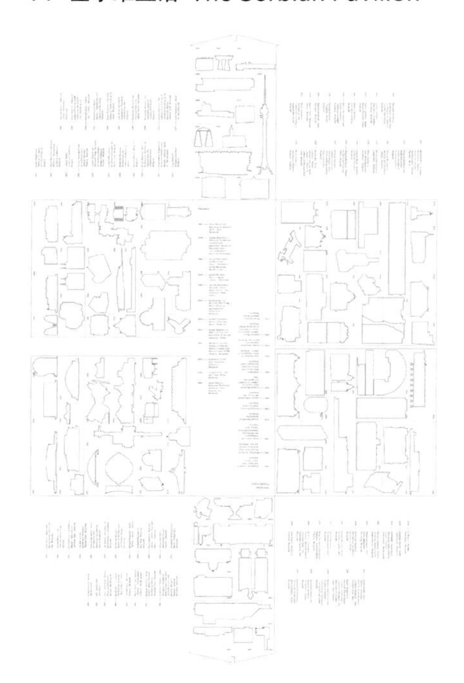

14–14

塞尔维亚馆将通过14–14的装置来展现一个世纪以来对建筑变化的审视。这种设计来自于一项对百年建筑作品的研究结果。此次展览的目的在于展示发生于我们社会和其建筑方面变化的影响。展览由两部分组成，包括100件作品和一座变革博物馆，是五位作者，即五位建筑师的作品。

100件作品

1914年至2014年之间创作于塞尔维亚的100件建筑作品展示在中心位置。选择这些作品是依据每一项目建造时人们对其产生的好评，它在一定程度上说明我们的社会和建筑专业有多么重视当代的建筑作品。

中央房间内的五面墙剖面都承载了一个构件，该构件明显突出，并说明作品中的问题。选取的五个基本构件包括：柱体、墙面、洞口、垂直连接和屋顶。新形成的秩序打破年代表和参观者的轨迹，使参观者在其中探寻特殊性与相似性、地方性与国际化、普通性与非凡性。

14-14

The Serbian pavilion will examine a century of changes represented in the installation 14-14. This comes as a result of a research into architectural production spanning one hundred years. The exhibition aims to show the effects of changes that happened in our society and its architecture. The exhibition comprises two parts, 100 works and The Museum of the Revolution and is the work of five authors, architects.

100 works

A hundred works of architecture, made in Serbia between 1914 and 2014, are presented in the central room. The selection is based on the critical acclaim at the time, and each project was made illustrating to an extent how our society and architectural profession valued its contemporary production.

Each of the five wall sections of the central room hosts an element, which stands out and defines the works in question. The five fundamental elements chosen are: the column, the wall, the openings, vertical connections and the roof. The newly formed order breaks with chronology and orbits around the beholders to trace out the specific and the similar, the vernacular and the international, the ordinary and the extraordinary.

照片提供：©Diego Terna

照片提供：courtesy of the Moroccan Pavilion (©Luc Boegly)

>> 摩洛哥馆 The Moroccan Pavilion

原教旨主义者

受到地域的启发，摩洛哥馆探索激进的和试验性的研究方法。除了欢迎国际建筑师的作品之外，摩洛哥首先是一片具有探索精神的土地，它是名副其实的先进项目的实验室。

摩洛哥鼓励构造和材料方面、形式和建构方面以及家庭和社会方面独特的建筑实验，这些实验为建筑历史做出了实实在在的贡献。而国家极其特殊的历史条件促成了这种发展，摩洛哥人的建筑设计天赋也包括吸收、消化和最终同化先进项目的能力。也许这也是摩洛哥传统（现代性和激进性传统，也包括专用和整合传统）的一部分。为了陈述这种冒险精神，摩洛哥馆计划设计一次侧重于关注居住基本问题的展览，并且利用同时期对沙漠中适宜居住的构筑物的思索将城市住宅的历史轨迹与前者并置。

为创造这种经历，整体的展示设计方案基于三个要素。地面：该馆全部200m²的地面都铺有沙漠的沙子。天空：一个120m²的屏幕仿佛天穹一般悬于兵工厂的结构之上。碑石：每一个项目占据一个1m³的虚拟体积，展示于具有不同高度的、1m²的碑石之上，而这些碑石从沙石中显现出来并按照规则且连续的网格顺序排列，使人联想到地理无穷大的形态。

Fundamental(ism)s

The Pavilion of Morocco explores the radical and experimental approaches inspired by the region. In addition to having welcomed the work of international architects, Morocco has above all been a land of exploration, a veritable laboratory for the Modern Project.

Morocco encouraged unique architectural experiments – constructive and material, formal and architectonic, but also domestic and social – that tangibly contributed to the history of architecture. While the country's very specific historical conditions fostered this development, the Moroccan genius has also consisted in the ability to absorb, digest, and ultimately metabolize the modern project. Perhaps this, too, is part of the Moroccan tradition – a tradition of modernity and radicality, but also of appropriation and integration. To narrate this adventure, the Pavilion of Morocco proposes an exhibition that focuses on the fundamental question of habitation and juxtaposes a historical trajectory of urban housing with contemporary speculations on inhabitable structures in the desert.

In order to create this experience, the overall exhibition design scheme is based upon three elements. The Ground: All 200 square meters of the pavilion are covered in desert sand. The sky: Like a celestial vault, a 120 square meters screen is suspended from the structure of the Arsenale. The steles: Each project occupies an imaginary volume of one cubic meter, displayed on one-square-meter steles of various heights, which emerge from the sand and are arranged according to a regular and continuous grid that evokes a form of geographic infinity.

>> 巴林馆 The Bahrain Pavilion

原教旨主义者和其他阿拉伯现代主义

原教旨主义者和其他阿拉伯现代主义是对殖民主义、原教旨主义以及阿拉伯世界现代性之间联系的一种探索，它试图理解这些运动之间的联系以及它们与当时的建筑学之间的关系。展览由一个大型的阿拉伯世界地图组成，它是以殖民时期的地图作为参考。一个圆形图书馆包含展品目录，并在地图周边集合了100个1914年至2014年期间建造于阿拉伯的项目。在该装置的上方，一个圆屋顶为Safar工作室受托制作的电影剧本提供投影，投影诠释了22个阿拉伯国家的国歌。七篇不同的短文构成了此次展览的系统内容，并对阿拉伯世界的不同地域（伊拉克、阿拉伯半岛、埃及、黎凡特、北非和中非）内的建筑发展进行了一次历史性的回顾。每篇短文都伴有收集的该地区1914年至2014年历史时期内建筑的相关档案文件。这些短文和收集的100栋建筑将构成标题为《1914至2014年阿拉伯世界建筑》展品目录的基础，参观者可在场馆内进行相关咨询。

Fundamentalists and Other Arab Modernisms

Fundamentalists and Other Arab Modernisms are an exploration of the links between colonialism, fundamentalism and modernity across the Arab World, in an efforts to understand the links between each of these movements and their relations to the architecture of the time. The exhibition consists of a large-scale map of the Arab World that references the colonial maps. A circular library that contains the exhibition catalogue and gathers a selection of a hundred built projects between 1914 and 2014 across the Arab World surrounds the map. Above the installation, a dome supports projections of a commissioned screenplay by Studio Safar which comprises of a reading of the 22 national anthems of the Arab countries. A collection of seven different essays forms the scientific content of the exhibition and offers a historical overview of the evolution of architecture within the different geographical components of the Arab World: Iraq, the Arabian Peninsula, Egypt, the Levant, North Africa and East Africa. Each essay is accompanied by a selection of relevant archival documents of buildings from that region within the historical period of 1914 to 2014. The collection of essays as well as a selection of a 100 buildings will form the basis of the exhibition catalogue entitled, *Architecture from the Arab World* 1914-2014, available for consultation at the pavilion.

照片提供：©Diego Terna

©Tobias Madörin

>>96

Herzog & de Meuron

Was established in Basel, 1978. Has been operated by senior partners; Christine Binswanger, Ascan Mergenthaler and Stefan Marbach, with founding partners Pierre de Meuron and Jacques Herzog from the left. Has designed a wide range of projects from the small scale of a private home to the large scale of urban design. While many of their projects are highly recognized public facilities, such as their stadiums and museums, they have also completed several distinguished private projects including apartment buildings, offices and factories. The practice has been awarded numerous prizes including "The Pritzker Architecture Prize" in 2001, the "RIBA Royal Gold Medal" in 2007.

>>146

Rafael de La-Hoz Arquitectos

Rafael de La-Hoz was born in Spain in 1955. Graduated from the Higher Technical College of Architecture of Madrid, and obtained an MDI Master in the Polytechnic University of Madrid. Now he directs his own architectural studio and works in Spain, Portugal, Poland, Romania, Hungary, and the United Arab Emirates as well.
Is a visiting professor at several universities, and is a member of the Editorial Council of the COAM Architectural Magazine.

>>114

Triarch Studio

Since graduated in 1972 from the National University of Architecture of Venezia(IUAV), Walter Giovagnoli has taken part in rebuilding and conversion programs, and furnished various hotels and entertainment venues. He was responsible for the restoration of a historical heritage church in Verucchio, Italy. He is now focused on one-family houses, farm stay properties, and commercial buildings.
Alessandro Quadrelli, graduated in 2004 from University of Ferrara, has worked with many firms in Emilia Romagna. Since 2006 he has been collaborating with Walter Giovagnoli in Triarch Studio. He is specialized in residential and commercial planning and construction, and interior fit outs. Now he lives in New Zealand.

>>104

Coarchitecture

Hudon Julien Associes was founded by Michel Hudon and Denis Julien in 1976. Have been specializing in architecture and urban design in Quebec. Is now headed by four emerging architects–Normand Hudon, Marie Chantal Croft, Alain Tousignant, Cesar Herrera and Louis Caron, the senior architectural technician. To develop a meaningful, innovative and sustainable architecture, they try to communicate, coordinate and control the risks. For the customers and their community, they devote their efforts to accelerate the transition to a sustainable development of our country. Without compromising, sustainable development and quality designs then become their priorities and it made the renown of the firm.

>>124

Branch Studio Architects

Is an award-winning Melbourne based design-intensive, architecture and design practice specializing in: architecture, interior architecture and design, urban design and master planning, landscape architecture and off branch custom furniture, one-off joinery pieces and art based installations. They work across a wide-range of project types and scales ranging from residential alterations and additions, new residential, multi-residential, commercial, health care, hospitality and retail, educational, liturgical, cultural and civic projects. It is their belief that architecture is far more enriching when it has meaning and substance linking each project to site, context, culture of place and inhabitants. Brad Wray is a director and architect of Branch Studio Architects. He completed his architectural studies at Royal Melbourne Institute of Technology(RMIT) in 2007 completing a Bachelor of Architecture with first class honors.

>>160

Lake Flato Architects

Was established in 1984. Was honored with the Firm of the Year Award from The American Institute of Architects in 2004, and with a Texas Medal of Arts in 2009. As architects, teachers, environmental stewards, and community advocates, they strive to elevate the public appreciation of architecture and foster the education of the next generation of architects. They also believe that architecture and sustainability are inseparable, and that buildings should be beautiful, affordable and successfully merged with the landscape as well as promote healthy living.

RSP Architects

Beau Dromiack received B.S. in Design Studies and M. Arch from Arizona State University. His recent work has been dedicated to higher education projects. Leads every project with a passion for not only meeting the needs of the client, but taking care to recognize the ultimate users.

>>54

Mikou Design Studio

Is a place of creation and experimentation in architecture and its inter-disciplinary cross-fertilization. Is currently involved in a large number of projects throughout France, Germany, Brazil and Morocco. Two partners, Salwa right and Selma Mikou left were born in Fes, Morocco in 1975. After attending school in Paris Belleville and EPEL, they received a diploma on Architecture and Urban Design in 2000.

>>36

Skylab Architecture

Is a Portland-based design studio practicing contemporary architecture, interior and branded design lead-ed by Jeff Kovel(principal architect) and Brent Grubb(principal). They are a team of professionals research-ing unique, original environments. The firm's work spans the industries of hospitality, retail branded envi-ronments, residential and commercial. Award-winning works include the Columbia Boulevard Wastewater Treatment Engineering Building, Nike Camp Victory, the W Seattle, and best of year recogni-tion for two commercial projects: the Flavor Paper Headquarters in Brooklyn, and the North office in Portland.

>>136

Archea Associati

Architects Laura Andreini, Marco Casamonti and Giovanni Polazzi were all born in Florence and trained at the School of Architecture of Florence. In 1988, they founded the Studio Archea based in their home-town. In 1999, architect Silvia Fabi also joined the office. In 2001, branches in Rome and Genoa were opened with the participation of Gianna Parisse and Massimiliano Giberti. Besides the main activities of the office which consist of research on design and implementation of architecture at various scales, members of the Archea Associati reconcile intensive occupations of teachers and researchers in several universities in Italy within the area of architectural design.

>>46

HHS Planer+Architekten AG

Manfred Hegge, a representative of the company was born in 1946 in Korschenbroich, Germany. He majored in Architecture at Ulm School of Design in the University of Stuttgart. Also studied Systems Engineering from Technical University of Berlin(TU Berlin) and Planning Studies from London School of Economics and Political Science(LSE). Has been teaching at many educational institutions including University of Stuttgart, University of Kassel, University of Hanover and Technical University of Darmstadt. He is also a member of the Federation of German Architects, Association of German Architects(BDA).

>>20

Tectoniques Architects

Joanna Relander graduated from the School of Architecture Paris Val de Marne in 2002. She's been an architect of Tectoniques since 2007. With 20 years of professional experience, Tectoniques is collective in structure, comprising two generations of architects and engineers. Specializing in dry construction tech-niques, Tectoniques has for long been interested in the principles associated with timber frame buildings. The firm is active in numerous sectors of architecture, including public buildings, housing and leisure facilities, with a strong commitment to environmental standards. Tectoniques applies construction proto-cols that are at once simple, clean and progressive.

>>66

Szyszkowitz-Kowalski+Partner ZT GmbH

Is an Austrian-German architectural design team founded by Karla Kowalski and Michael Szyszkowitz in 1978. They put great emphasis on three-dimensional and expressive architec-tural language with a distinctive reference to landscape and context. Michael Szyszkowitz was born in 1944, Graz studied at Graz Technical University. Since 1998, he has been working as a professor and the head of the Institute for Building Construction and Design at Braunschweig Technical University. Karla Kowalski was born in 1941, Beuthen. She studied at Darmstadt Technical University. Since 1988, she has been a professor and the head of the Institute for Public Buildings and Design at Stuttgart Technical University.

Julian Lindley
His interest in research and design revolves around two possibilities. Firstly, exploring new approaches within design methodology and education. That is how will this new technology alter the way we develop products and how do we respond creatively. Secondly, how do we utilize different design approaches in creating a sustainable future. Within this he acknowledges the paradox that currently design has increased rather than decreased material consumption. Has lectured internationally on his main research interests, the theme of Design Methodology and Sustainability.

Fabrizio Aimar
After graduating cum laude from the Polytechnic University of Turin(Italy), Fabrizio has worked the past five years for a civil and infrastructural engineering office in Turin, where he has developed a variety of structural projects in collaboration with such international architectural firms as Jean Nouvel, Renzo Piano, Mario Cucinella and Aymeric Zublena. He has been a contributor since 2010 to the Italian architectural magazine Il Giornale dell'Architettura. He has been a member of the Cultural Committee of the Asti local board of Architects since 2010. Is currently working on the new Intesa Sanpaolo skyscraper in Turin, designed by Renzo Piano.

Simone Corda
Is an architect and Ph.D. candidate based in Sydney, Australia. Explores the themes of contemporary architecture through researches and projects at different scales and cross sectors. Referring to the architecture of Australia and New Zealand, he is currently focusing on the flexibility of housing as the key concept for sustainability. Part of his PhD thesis about Glenn Murcutt's work has been already published in the Italian magazine Area. Contributes to the Faculty of Architecture at the University of Cagliari enthusiatically with regular seminars and lectures at the Faculty of Architecture in Alghero, C.N.R. National Center of Research and Festarch event.

©Stephano Goldberg

>>76

Renzo Piano Building Workshop
While studying at Politecnico of Milan University, Renzo Piano worked in the office of Franco Albini. After graduating in 1964, he started experimenting with light, mobile,temporary structures. Between 1965, and 1970, he went on a number of trips to discover Great Britain and the United States. In 1971, he set up the Piano & Rogers office in London together with Richard Rogers. From the early 1970s to the 1990s, he worked with the engineer Peter Rice. Renzo Piano Building Workshop was established with 150 staff in Paris, Genoa, and New York.

图书在版编目(CIP)数据

节能与可持续性 : 汉英对照 / 韩国C3出版公社编 ; 张琳娜, 周一译. —大连: 大连理工大学出版社, 2014.9
(C3建筑立场系列丛书)
书名原文: C3 Energy Efficient,Sustainable
ISBN 978-7-5611-9542-0

I. ①节… II. ①韩… ②张… ③周… III. ①节能—建筑设计—汉、英②建筑设计—可持续性发展—汉、英 IV. ①TU2

中国版本图书馆CIP数据核字(2014)第216398号

出版发行：大连理工大学出版社
(地址：大连市软件园路 80 号 邮编：116023)
印 刷：上海锦良印刷厂
幅面尺寸：225mm×300mm
印 张：12
出版时间：2014 年 9 月第 1 版
印刷时间：2014 年 9 月第 1 次印刷
出 版 人：金英伟
统 筹：房 磊
责任编辑：张昕焱
封面设计：王志峰
责任校对：赵姗姗

书 号：978-7-5611-9542-0
定 价：228.00 元

发 行：0411-84708842
传 真：0411-84701466
E-mail: dutp@dutp.cn
URL: http://www.dutp.cn